The GOODE GUIDE *to* WINE

The publisher and the University of California Press Foundation gratefully acknowledge the generous support of the Simpson Imprint in Humanities.

SIMPSON

IMPRINT IN HUMANITIES

The humanities endowment

by Sharon Hanley Simpson

and Barclay Simpson honors

MURIEL CARTER HANLEY

whose intellect and sensitivity

have enriched the many lives

that she has touched.

The GOODE GUIDE *to* WINE

A Manifesto of Sorts

JAMIE GOODE

UNIVERSITY OF CALIFORNIA PRESS

University of California Press
Oakland, California

© 2020 by Jamie Goode

LIBRARY OF CONGRESS CATALOGING-IN-PUBLICATION DATA
Names: Goode, Jamie, author.
Title: The goode guide to wine : a manifesto of sorts /
    Jamie Goode.
Description: Oakland, California : University
    of California Press, [2020] |
    Includes index.
Identifiers: LCCN 2020000759 (print) |
LCCN 2020000760 (ebook) |
ISBN 9780520342460 (cloth) |
ISBN 9780520974616 (epub)
Subjects: LCSH: Wine tasting. | Wine and wine making.
    | Wine selection.
Classification: LCC TP548.5.A5 G664 2020 (print) |
    LCC TP548.5.A5 (ebook) | DDC 641.2/2—dc23
LC record available at https://lccn.loc.gov/2020000759
LC ebook record available at https://lccn.loc
    .gov/2020000760

29 28 27 26 25 24 23 22 21 20
10 9 8 7 6 5 4 3 2 1

Contents

Preface

Herewith: a wine manifesto. Of sorts.

WHY?

It's an attempt to gather together some of my thoughts about wine, in a series of short, targeted chapters.

For the last few years I have been spending most of my time on the road, visiting wine regions, meeting wine producers, taking part in conferences, and tasting, tasting, tasting. I have even spent time in wineries during vintage. I've learned a great deal, but because wine is such a complicated, fragmented topic, I suspect a lifetime of travel is needed to understand this world fully.

Like any sphere of human activity, there is good and bad in the world of wine, and a whole lot of stuff in between. There are many segments in the market, and tens of thousands of different wines. Each vintage is different. This creates a bewildering matrix, and to make sense of it is far from straightforward.

My approach to wine is a bit different, and I think it is quite unique. And worth reading. If I didn't, I wouldn't waste my time in writing this book, and expect you to waste your time reading it. I was trained as a scientist, but I'm an artist at heart. I have a PhD in plant biology, and I worked as a science editor for fifteen years before I gate-crashed the wine world. But while I'm a scientist, I'm not a scientific fundamentalist. I acknowledge that there are different ways of seeing the world, and while science is useful—essential, even—we need to acknowledge that a purely scientific narrative gives us an incomplete story. In life and in wine.

Where should wine be heading? I make some suggestions.

I look at the way we perceive wine. I ask: Is wine special, and why? I poke (good-natured) fun at some of the sillier aspects of the world of wine. And where necessary, I warn. But all of this, of course, is from my perspective. I don't expect everyone to agree.

In essence, this book provides a philosophical underpinning for fine wine. It is our philosophy that shapes how we act; yet we rarely verbalize these principles and theoretical insights.

This book is intended for a broad spectrum of readers, including interested consumers and those in the wine trade—whether you are drinking, making, selling, or marketing wine, there are chapters specifically addressed to you. The first half of the book is aimed more at the wine drinker; the latter half has ideas relevant to those in the trade.

I will repeat myself: you may disagree with me. That's OK. But I hope the ideas presented here cause you to question, and perhaps in turn to frame your own philosophy of wine.

The heart of authenticity

*Authentic wine is rooted in a place
and a time.*

WINE BELONGS IN the same family of foodstuffs as beer, cheese, coffee, and even chocolate, where microbes play an essential role in the production process. But there's something unique about wine, in that it is a product of a particular time and place, both of which write themselves on the sensory properties of the resulting beverage. You could argue that cheese is similar in this regard: the season will affect the way the grass grows, and this will affect the taste of the cheese by altering the properties of the milk. And of course, different places produce cheeses that have a local taste. But wine—decent wine, that is—takes this quality further.

The special thing about wine is its connection to a particular place, and to a particular time. Of course, not all wines have this connection. Some very fine wines are a blend of vintages, as are some very cheap ones. And some wines are made in such a way that it's very hard to know where they came from. But the important fact here is that for the sorts of wines that I usually like to drink, and which I enjoy the most, there is a definite connection between the place and the wine: a local flavor, known widely as *terroir*.

Much has been written about terroir. All I want to say here, for now, is that it's the defining concept in fine wine. If you grow a grape variety (or varieties) in the appropriate place for these varieties, and your viticulture is good, and then you don't mess things up too much in the winery, you will have a wine that tastes of its place. Some places have more personality than others, of course: not all terroirs are created equal. The proof of terroir, though, lies in the hands of winegrowers who make wines the same way from different plots, and then the resulting wines show differences that can only be due to the site. That's the proof of a concept, and from that starting point we can keep going, expanding this concept and making it more nuanced. It's what makes wine so interesting.

Do all wines have to be terroir wines? No: I think there is a place for multi-site, or multi-region, blends. Vintage Port is the result of skilled blending, and the top Champagnes are also frequently from multiple vineyards. You could argue, however, that the skilled blenders are working with good terroirs, and understand terroir. It creates the components for successful blending. There's also the famous example of Penfolds Grange, Australia's most celebrated fine wine and a multi-site and multi-region blend. I would argue, though, that these exceptions don't call into question my assertion that terroir is at the heart of fine wine.

And what of time? A wine is not fixed. It is born in a growing season, and bears the characteristics of the weather of that year. The matrix of vintage and place is infinite, yet place should be the anchoring point. Vintage, then, becomes the lens through which we see place, and follow it over time. The wine develops in the bottle until, eventually, the signature of place is lost and the wine grows timeless and old. No wine is infinite, and this temporal quality reflects our own mortality.

The skill of winegrowing

*Sensitive, intelligent winegrowing[1]
produces wines that capture the location
and the vintage.*

WINEGROWERS, WHEN QUESTIONED, often talk about
being custodians of their vineyards, saying that they
don't make the wine: it is the vineyard that produces
the wine. They say that they do nothing in the winery.

But to do nothing requires great skill if the result
is going to be interesting. Doing nothing is an active
choice when it comes to proper winegrowing. It relies

---

1. A note on language, for language is important. It shapes our perception. Careful use of language reminds us of important issues. For this reason, I believe that we should stop using the term *winemaker*. My preference is for *winegrower*. It better reflects the role of human agency in the production of wines, which at its heart is a microbiological transformation.

on understanding and observation, not sitting back and hoping for the best.

Terroir often speaks with a quiet voice, and winemaking interventions can drown out the vital, nuanced signatures of place, and flatten the differences among vintages.

Sometimes, though, intervention is needed, to capture terroir and to bring out one of the most interesting facets of wine: vintage personality. Think of winegrowing as an act of interpretation. If you are a skilled, experienced grower, you have an idea in your head of how your sites should express themselves in terms of wine flavor. Flavor is an abstraction of place: it is quite a jump from a place to a liquid in a glass. So the art of winegrowing is to produce an intelligent interpretation of place.

If we think of the terroir as a radio signal, then clearly you want to tune your radio to pick up that signal as strongly as possible. If the winegrower's hand doesn't have access to the tuning knob, the signal can drift, and the result will be noise. In this case, we are considering terroir to be something real— something that can be lost or found. I think it is, and I like the idea that winemaking is about retuning, to stay with the signal and thus to hear the place as clearly as possible. That is what I mean by appropriate intervention.

Then we have the vintage. In the past, great vintages were rare, and even good vintages weren't the norm. Now, advances in winegrowing have meant that poor vintages are less common, and good vintages almost the norm. The vintage signature, to a degree, has been muted. Is this an entirely good thing? No one wants a bad vintage, and we regret them when they arrive. Yet some vintages surprise us. Although we may have discounted them or written them off at first, in time they took a turn for the better. Perhaps we should be humbler in the face of vintage variation, and embrace it more fully? Thus the winegrower has a twofold duty: to strive to interpret the place, and to read the vintage in an intelligent way.

Soils matter

The ceiling for wine quality is determined by the soil. Great wines can only be made from privileged terroirs, no matter how skilled the winegrower and how perfect the climate.

IT ISN'T EASY to understand the relationship between wine and the soil. Intuitively, we see the vine rooted in the soil, and each season there's new growth: the canopy expands, and the grapes form and ripen, ultimately, via fermentation, giving flavor to the wine. It seems that all this is coming from the soil.

Not so, say the scientists. All the vine is taking up from the soil is water and mineral ions. The vine, like all plants, adds sunlight from the sky and gases from the air: acting as a chemical factory, it uses photosynthesis to fashion all manner of complex chemicals. What a miracle!

Somehow, though, this grapevine physiology is sensitive to place. Small variations between one vineyard and another create grapes whose chemical composition differs slightly. Then, after the process of fermentation and maturation, these initial differences result in wines that we can tell apart.

Let's explore the scale of these differences. Within each bunch, individual grapes will differ, and there will be differences from vine to vine. This scale is too small for us to bother with, though; instead we must approach winegrowing at a coarser level—that of vineyard blocks. Here we are looking at collective differences in the grapes that are harvested from these blocks, and that are then selected (if a sorting table is used) to go into the vat.

Clearly, the microclimate has some effect on the way the grapes develop. Light is important, as are temperature, wind, and rain. Different vineyard blocks will experience different microclimates, and through the physiology of the vine this will affect the grape berry composition. But what of soils? They seem to matter, too—but how much and exactly how?

The scientists tend to ignore the influence of soil chemistry. They think that the influence of the soil is largely through its water-holding capacity, and the way in which it then delivers water to the roots of the vine. Yet many of us love to talk about how wines grown on granite taste different from those on limestone or schist, or we think about the role of clay (and the different sorts of clays) in the flavor of the wine. Then we have the influence of the soil life on the way the vine grows. Indeed, this is an area of intense investigation at the moment. All sorts of signaling takes place underground between the vine roots and the microbes that surround them, and it's intriguing to think that this could, albeit indirectly, be affecting wine quality.

Although we lack the facts to support our assertions, many of us who have seen lots of vineyards and tasted wines from them critically, think that it's soils that determine just how good your wines can be. It's clearly not just the climate. The proximity of great and mediocre vineyards makes this clear. Their climates will be very similar—and remember, a vine never sees the climate: it sees the weather of the year. A climate is just an average. There will be more variation in weather from year to year than there will be in the climate experienced by two closely located vineyards in any one season. So the soils are having an effect.

The arrogance of the new world is the belief that if you have the best viticulturists and the best winemakers, you can make great wine. This ignores the wisdom of many centuries of winegrowing in the classic wine areas. The new world experts look at the restrictions placed on old world winegrowers—where they can plant, what they can plant, how they grow their vines—and laugh. They dismiss terroir as a marketing ploy. They plant new wine regions free of any restrictions, matching grape variety to site by looking mainly at the climatic conditions and pretty much ignoring the soil. Then thirty years on they slowly come to the realization that the old world winegrowers aren't entirely stupid. Soils matter, and some vineyard areas are more privileged than others. The restrictions of old world appellations, it turns out, are just official sanctioning of centuries of experience: some places are better than others, and some vineyard regions have talent for certain varieties.

Now in the new world, the best sites are becoming apparent. The search is on for great vineyards, and to find them requires surveys and mapping—and digging lots of backhoe pits. Not all vineyards are created equal, and great wines can only be made from great vineyards. Soils matter.

The art of interpretation

*A particular terroir can have a number
of interpretations. It's wrong to think
that there can be just one wine made
from each site.*

IF YOU ARE a musician looking to cover a well-known
song, it's wise to do your own interpretation, in your
own style, playing to your particular strengths. As
most of your audience will have heard the famous
song many times, your mimicking better be good. And
there's a chance they might already be bored with the
original, if it is very famous. To breathe life into it, you
need to add something of yourself.

It's a stretch to apply this analogy to wine, I
guess, but I'm going to anyway. My point is that
there can be many different versions of the same
wine made from a specific vineyard site, or region,
or grape variety, and among these many versions

there will be some that are better and some that are worse.

It's really interesting to see several wines of the same vintage from the same vineyard side by side. If this is a worthwhile vineyard, you'd expect there to be some family resemblance among the wines, but you wouldn't expect them all to taste the same. In one wine, you might find the signal of the vineyard diluted by some prominent new oak; in another, you might find a trace of *Brettanomyces* adding animal notes that compete a bit for attention. But for the rest, perhaps the character will be clearly recognizable, played slightly differently in each case. It is for us as the drinker to choose which interpretation we prefer—and after that it's a matter of discussion to decide which we personally believe to be the truest.

Recognizing the local flavor of a wine to be an act of interpretation means that we have brought the human element into our definition of terroir. This is as it should be. Because of the leap from the growing of grapes in a vineyard to the perception of flavors in a wine glass, terroir has to be partly cultural. Ultimately, it is for us to decide together about legitimate and illegitimate expressions of a terroir, and this will depend on many factors. The traditional winemaking practices in a region may be reflected in the way that the wines from that place are recognized to taste. Ultimately, any winemaking step that obscures site differences is going to be problematic, even if it is rooted in tradition and makes wines from an area taste the way they do.

Some wines are just wine

Of many wines there is nothing to be said. They are just wine. It's foolish to say anything more about them, but still some people try.

THE WORLD NEEDS good cheap wine. When I was a kid in the late 1970s, we used to go camping in the south of France and Spain, and there was a lot of cheap wine around. Much of it never saw a bottle. Some of it was sold in simple one-liter plastic bottles, like water. This wasn't wine that was going to hang around for a long time. Occasionally my parents would go to a bodega. There, you took your own container, filling it straight from the tank. Inexpensive wine was a commodity, not something to linger over—more of a staple.

In southern European countries until fairly recently, for most people wine was food. It was drunk in large quantities daily, not with the goal of drunkenness, but to slake thirst and to accompany meals. In Portugal, workers would take wine with them into the fields. I'm sure it was the same in many rural parts of Europe where wine was grown. Getting boozed-as was not the aim.

These wines are just wine, and there's an honesty to them. They are dying out, however. Now we live in cities and we want wine to taste nice. Wine has become fancy and aspirational. The plonk of the past didn't taste nice; it just tasted like wine.

Why can't we be comfortable with wine as wine? The integral truth of humble commodity wine has largely been lost. Wine is suddenly trying to be oh-so-special—even cheap wine. It is dressed up in wine-making trickery, packaged to look more expensive, with bold marketing claims on the back label. These wines are borrowing their identity from fine wine, but that's not what they are. They are imposters, liquid confidence-tricksters. Wine critics write glowing tasting notes on them, but of many of these wines, in fact, there is simply nothing to be said.

Why do we lose our honesty when it comes to cheap wine? What is it about wine that makes us behave differently? It's like the artificial, strangled reverence shown by a congregant in a traditional church service. It's as if we've somehow forgotten that in order to make sense of wine, we need to segment the marketplace, and acknowledge that there's such a thing as commodity wine, and there's such a thing as fine wine, and there are some layers sandwiched in between. The rules are different at different levels.

Wine can be simply wine, a food, something that just needs drinking, and this can be pleasurable and life-affirming. If, that is, it is honest wine. Let commodity wines be what they are. We don't even need to turn the experience into words; we can just dwell in the experience, and that's OK.

6

The wine is a whole

Reductionistic approaches to understanding wine—breaking it up into its various components—have a place and some utility. But this utility is limited. If we want to understand wine properly, we need to take a holistic approach.

SOME PROBLEMS ARE very big indeed. How do we solve them? We break them down into smaller pieces, and then study those more manageable pieces. Then we bring the resulting knowledge together and hope that our modular approach has provided us with a global level of understanding. This is the way most science is done, and it's known as reductionism. It can be a pretty powerful approach.

Back in the day, we didn't know how the brain worked. So people removed brains from cadavers and looked at them with the tools they had available. They cut sections and looked at the sections under microscopes. They used dyes and stains. They got better microscopes that enabled them to see in finer detail. They started looking at patterns of electrical activity in live brains. And they learned from disease and dysfunction: examining the effects of damage to specific regions of the brain can provide insight into these parts' role during normal function. And then they developed magnetic resonance imaging (MRI), which allowed them to see the brain light up while it is performing specific tasks. This reductionistic approach has given us a lot of understanding of brain function.

Yet still we can't explain consciousness. Or the human personality. Or the soul. There are limits to reductionism.

When you begin learning about wine, you are taught how to write structured tasting notes. These are highly reductionist in nature. Focusing one by one on the disparate elements of the wine, you describe them. What does it look like? How does it smell? How does it taste in the mouth? What is the finish? You are translating your perception into words, hoping that this reductionist tasting note captures the wine in a meaningful way.

Or you might be judging a competition where you are scoring on a twenty-point scale. Again, you break the wine up into its components, awarding scores for appearance, nose, palate, and perhaps a few points for overall impression. Yet there is limited utility in this sort of approach.

The wine is a whole, and we perceive the wine as a whole. Our brains are taking in various sensory inputs and processing them all, delivering to our consciousness a perception that is based on the liquid in the glass, but which has been edited. How do we describe this wine?

The very questions we ask of the wine will in part shape the answers we arrive at. Are we focusing too heavily on components, rather than on the overall wine? If we rush too quickly to words, we will miss out on the experience. We need to listen to the wine. Indeed, the words that we possess for the wine will probably shape our perception of it. If we are not careful, we can taste without tasting.

Reductionism is important in our efforts to learn about wine, but there comes a point where we need to think more holistically. We need to see the wine as a whole. We need to dwell with it, experience it, and enter into a creative process with a to-and-fro approach. Our perception is an interpretation.

And let's *consider* tasting notes, while we are sort of skirting around the subject. They are mostly terrible. I really hate them. Our capacity to express our experience of wine in words is very limited, and if we look back to wine writing of fifty years ago, the tasting note as we now know it didn't exist. Today we are bombarded with a legion of wine critics, following in the wake of the immensely successful Robert Parker. They churn out mostly hideous tasting notes and anoint wines with (mostly) absurdly high scores, in their clamor to be quoted by wineries and sell score stickers and entries at their wine events. Yes, we do need to try to describe wines in our work as wine writers, but I would just love to see better, more holistic, more intelligent tasting notes. It's something I aspire to, though it's very hard to do.

7

Wine resists the proud

*Be humble in the face of wine. It is an
endlessly complex subject, changing with
each vintage. Indeed, wine is beyond any
single human's ability to understand to
any serious degree. We see only in part.
And that's OK.*

WINE CAN MAKE you look stupid, and quickly. Especially when you are tasting blind, which is a satisfyingly humbling experience—at least for those watching. There is an imprecision to wine tasting. Even when we are concentrating hard in our interrogation, sniffing and slurping, there's often the sense that the wine is yielding up only part of its essence. I don't think I'm all that bad at blind tasting, but my triumphs, where I spot what the wine is, are countered by the many times where I am left embarrassingly clueless. You have to feel for the candidates of the Master Sommelier exam, who have to identify six wines blind to pass their practical exam. Just six!

And then there's the incredible breadth and complexity of the world of wine. An array of places, soil types, slopes, aspects, grape varieties, farming practices, harvest timings and techniques, winery practices, yeasts, bacteria, and blending decisions results in a bewildering array of flavors. There's the complexity of classifications, and also the magical element of time, expressed both in vintage variation and in the way a wine develops over time. This means that you can spend your professional life in wine and still have more to discover.

Many consider this a problem to be corrected. For a big company in charge of a wine brand, such variation is the enemy. They want to make things simpler. This complexity can also make it hard for the average consumer to choose wines—although to many, wine is simply a commodity: as long as it tastes OK and is affordable, it'll fit the bill.

Most non-involved consumers live quite happily with wine as a commodity, although many have the nagging thought in the back of their minds that they really should know a bit more about wine (which isn't true; it's entirely optional and not everyone is all that interested in wine), because they have heard that it's a complex subject and that you have to learn a lot to be able to choose wines wisely and get the most out of them (this bit *is* true).

It is the complexity of wine, however, that draws many people in. Demystify it at your peril if you wish to keep the keen consumer engaged. I think that many people can live quite well with a bit of mystery. It's the control-freak self-appointed consumer champions who want to iron out all the details and make it all much simpler. They should relax: wine can never be truly simplified without ripping out its heart. Indeed, we *all* need to relax a bit: wine is a complicated subject that is impossible to master. Even people with letters after their name don't know it all. It only leaves people confused if they feel they have to know all about it. We can certainly think about better ways of selling wine to non-involved consumers, but we should be very careful before we tear down what we now have—a compelling, complex world of diversity—to replace it with something easy to understand but largely not that interesting.

And those of us who work with wine must learn to live at peace with the complexity of this compelling liquid pleasure.

8

The taste is not in the wine

The taste of wine is not a property of the wine, but is a property of our interaction with the wine. We bring a lot to the wine-tasting experience.

HERE WE ENTER the worlds of philosophy, neuroscience, and psychology. Any concept that requires such a combination of disciplines may seem too esoteric a topic to be exploring, but I think it is important to consider, because it affects how we approach wine, how we talk about it, and even how we sell it.

As humans we have been shaped by evolution. The internal readout of the world around us—our conscious perception of the input from our senses—is cleverly designed to help us navigate the world successfully. There's stuff out there in the environment that we have no access to without specific tools. A good example is the electromagnetic wavelengths that our mobile telephones or radios can pick up. Humans navigate the world chiefly through our vision, while many animals map the environment through smells that we can't pick up. It follows that flavor—the multimodal sense combination of smell, taste, touch, vision, and even hearing—is adapted to our needs and is partial, in that there are many chemicals that have no strong smell or taste, while there are others that we are acutely sensitive to.

So flavor is a human perception. We have developed the ability to taste certain chemicals and combinations of chemicals, which then, after some brain processing, create a conscious experience. Wine has a chemical composition, which is a property of that wine. We experience a taste, which is created by the chemicals in the wine, in combination with our experience, expectations, and the context of consumption. So some of the flavor is coming from the wine, and some is coming from us and the environment it is consumed in. In this sense, the taste of wine is not a property of that wine, but of our interaction with the wine.

From a philosophical point of view, though, we now enter an interesting discussion. It revolves around whether we could consider the taste of the wine to be a property of the wine independently of our perception of it.

Imagine I am learning about wine, and I return to a certain wine a year later, after extensive tasting and study. The wine, which has been stored in a cool cellar under conditions where you would expect it to change little over twelve months, tastes different to me than it did before. My repeated experience with wine, and the experience and vocabulary I have gained over time, have changed my relationship with it. Yet the chemical composition of the wine in question has changed very little. So we can phrase this two ways: either the taste of the wine is different (because the taste is something within me, created by my brain, and it doesn't exist outside of a human observer), or I am experiencing the taste of the wine (as a property of the chemical composition of the wine) differently.

In this instance, the implication is that there are two different ways of looking at the taste of wine. The first is that wine has a chemical composition, and then we taste the wine, creating a perception that is subjective, because we are not measuring devices and this perception is in part dependent on us. The second is that wine has a chemical composition, which leads to it having a "taste." Here, it is our perception of this taste that is subjective.

The difference here is whether or not we invoke this intermediate stage in perception, which is the flavor of a wine. We have the wine's chemical composition, and we could argue that this is what gives the wine its flavor—i.e., that flavor is an objective property of the wine. It is our *perception* of this flavor that is subjective. This is quite tidy, and allows us to retain objectivity in wine appreciation: as tasters, we are coming to the wine with the goal of "getting" the wine. The better the tasting conditions, the better the temperature of the wine, the fewer distractions, the more we concentrate, the better our skill as a taster, the more likely it is we will perceive the flavor of the wine. Let's imagine that you are dining with a small group, and there are some smart wines on the table. As you drink, you discuss. Clearly, each of you will be biologically different, and will experience the wines differently. Yet the way you discuss assumes that you are experiencing more or less the same thing as you drink. You are subjective observers trying to get to something objective: the flavor of the wine.

If we omit this intermediate state—the flavor of wine—then our perception of wine is simply subjective, and what is objective is the chemical composition of the wine. It's not as tidy, because it leaves us swimming in a sea of subjectivity. The taste of the wine is really just something personal to the observer, created in their brain.

Wherever we sit on this subjectivity/objectivity debate, we come to the same conclusion. We bring our experience and background to the wine-tasting event. A novice and an experienced professional will approach a wine in different ways: they will ask different questions of the wine; the novice will have fewer reference points, and fewer expectations.

When we attempt to evaluate a wine, we might choose to write a short tasting note and assign a score. So often, we regard such descriptions to be a property of the wine. Yet really what we are describing is the perception we have when we interact with the wine. This note and score don't actually belong to the wine, but rather are the outcome of our interaction with the wine. They are based on the wine, and if we are experienced, and good at tasting, we might find others agreeing with us. But it's not as simple as us acting as machines—machines whose accuracy improves with experience and training. It's just not like that in reality. And that's probably a good thing.

9

Buildings, people, fabrics

We have an impoverished vocabulary for tastes and smells, so describing wine ends up a challenge. Figurative language is the best way of capturing the personality of wine in words. Shopping lists of ingredients are so inadequate.

TASTING NOTES ARE horrible and I hate them. Yet I have to write them. We experience wine, and then, because of the need to communicate, we have to translate these experiences into words.

This is something we do all the time. Words are an abstraction of experience. Language is part of being human, the ability to acquire language seemingly hard-wired into our brains. This raises an interesting question: Does our language, with its own specific vocabulary, in some way shape our experience? Do we use the lens of language as a way to see and make sense of the world around us? Translating this to wine, does our language for wine—the set of descriptors we have marshaled to create our tasting notes—affect our tasting experience?

These are good questions, but what I am concerned with here is the best way we can use words to capture a wine.

When I was new to wine, my tasting notes were absurdly simple. It was a frustrating experience, trying to convert my perception—which was detailed, rich, and multifaceted—into a tasting note. I just didn't have a good vocabulary for wine. So I began reading the tasting notes of others. I recognized the structure they used when they described wines, and began to learn the code. Some of these descriptors I recognized in the wine; others puzzled me.

There was a common theme to these notes, all of which read like ingredient lists. They were reductionistic, splitting wine up into its component parts. And they were somewhat alienating for someone like me, new to wine. They gave the impression that these tasters were discovering so much more in the wine than I was, an impression compounded by the certainty with which these notes were usually written. Still, after a while, my notes got longer and began to be filled with more descriptors. I had joined the club.

My sense of dissatisfaction persisted, however. What about the use of figurative language in tasting notes, which is much more holistic? Yes, we need descriptors, but we also need some more global terms that seek to capture wine as a whole.

As a colleague of mine, Chris Losh, who used to edit the drinks trade magazine *Imbibe,* observes, "Structured tasting notes are great if you are at a busy tasting with limited time, but they actively discourage originality. And by having a growing number of trade folk trained in this way, we are in danger of creating an industry full of well-educated, well-meaning automatons."[1] He recalls a session where he tasted whiskies with some bartenders. They'd never been educated about wine tasting, and wrote very different notes from the ones Losh came up with.

> Mine were solid, by-the-book—and desperately dull. Theirs were chaotic, erratic—and fizzing with inspiration. I still remember one of them describing a 12-year-old malt as being "like walking through a dew-covered garden on a spring morning"—a sentence that absolutely

1. Chris Losh, "Dear Wine Writer, What on Earth Are You Talking About?" Just Drinks, November 6, 2015 (www.just-drinks.com/comment/dear-wine-writer-what-on-earth-are-you-talking-about-comment_id118579.aspx).

nailed the product and also, crucially, made me want to drink it. And this, I think, is where the problem lies. Our structured approach to writing tasting notes is designed to strip visceral reaction out of the process—or at least to minimise it; to render our judgments coolly analytical and free of the distorting effects of emotion.

For all the progress that has been made possible by the structured approach to tasting, the mood in the wine world seems to be that we might have lost something. In 2015, noted American wine commentator Matt Kramer published a short book titled *True Taste: The Seven Essential Wine Words.* In it he prompts us to move away from a focus on flavor identification, to more global, thoughtful, subjective terms that capture the qualities of wine better. "Too many tasting notes now offer little more than a string of fanciful flavor descriptors with the judgment revealed only in the score itself—a numerical 'thank you ma'am' after the more energetic 'slam, bam' of the flavor descriptors." Kramer's seven words are: *insight, harmony, texture, layers, finesse, surprise,* and *nuance.*

Metaphors can bolster the impoverished language we have for describing tastes and smells, giving it new life.

Tasting notes employ various categories of metaphor: for example, wine as a living creature, wine as a piece of cloth, and wine as a building. While we'd like to have a more exact way of sharing our experience of wine in words, such precision doesn't exist, and those who restrict themselves merely to naming aromas and flavors end up missing out on some of the more important aspects of the character of wines, such as texture, structure, balance, and elegance.

So what is a good tasting note, and what is a bad one? It depends to a degree on the purpose of the notes. Are they attempting to describe a wine so that someone might be able to recognize it in a lineup on the basis of the note? Or are they trying to capture something more transcendent and emotional? With our notes, we are trying to express an experience and a feeling, not just describe the flavor molecules in the wine. But it is very hard to do.

Perhaps my favorite writing on this theme comes from Evelyn Waugh's novel *Brideshead Revisited,* set in the 1930s. In this lovely passage, the narrator, Charles Ryder, is recalling an idyllic summer he spent with his friend Sebastian Flyte at the latter's family home, Brideshead. There they discovered wine together. Surely, this is the way to write about the flavor of wine:

We had bottles brought up from every bin and it was during those tranquil evenings with Sebastian that I first made a serious acquaintance with wine and sowed the seed of that rich harvest which was to be my stay in many barren years. We would sit, he and I, in the Painted Parlour with three bottles open on the table and three glasses before each of us; Sebastian had found a book on wine tasting, and we followed its instructions in detail. We warmed the glass slightly at a candle, filled a third of it, swirled the wine round, nursed it in our hands, held it to the light, breathed it, sipped it, filled our mouths with it and rolled it over the tongue, ringing it on the palate like a coin on a counter, tilted our heads back and let it trickle down the throat. Then we talked of it and nibbled Bath Oliver biscuits and passed on to another wine; then back to the first, then on to another, until all three were in circulation and the order of the glasses got confused and we fell out over which was which and we passed the glasses to and fro between us until there were six glasses some of them with mixed wines in them which we had filled from the wrong bottle, till we were obliged to start again with three clean glasses each, and the bottles were empty and our praise of them wilder and more exotic.

"It is a little shy wine like a gazelle."

"Like a leprechaun."

"Dappled in a tapestry window."

"And this is a wise old wine."

"A prophet in a cave."

"And this is a necklace of pearls on a white neck."

"Like a swan."

"Like the last unicorn."

10

Framing: how words can get in the way

Words shape our wine drinking experience, but sometimes they can get in the way. By jumping straight into winespeak we can miss out on dwelling in the experience itself.

FRAMING IS A social science term referring to a set of concepts and perspectives that influence how we think on certain issues. In this sense, framing is part of the narrative structure with which we see the world. American cognitive linguist and philosopher George Lakoff popularized the term in his 2004 book *Don't Think of an Elephant,* about political discourse in the United States. For example, the term *tax relief* has a strong framing influence, as he explains:

> The word *relief* evokes a frame in which there is a blame-less Afflicted Person who we identify with and who has some Affliction, some pain or harm that is imposed by some external Cause-of-pain. . . . Relief is the taking away

of the pain or harm, and it is brought about by some Reliever-of-pain. . . . The Relief frame is an instance of a more general Rescue scenario, in which there is a Hero (the Reliever-of-pain), a Victim (the Afflicted), a Crime (the Affliction), a Villain (the Cause-of-affliction), and a Rescue (the Pain Relief). The Hero is inherently good, the Villain is evil, and the Victim after the Rescue owes gratitude to the Hero.

How does this apply to wine? The use of words is inseparable from our experience of wine. Even the names of wines, or of the grapes that they are made from, carry supplementary meaning—frames—that influence our experience of these wines. For example, I have a friend who claims that she hates Gewurztraminer. For her, the very word comes loaded with meaning, framing her experience of the wine negatively. If, however, she were to taste a Gewurztraminer blind and not realize what it was (a tough ask: after all, this is a very distinctive variety), there would be no such frame and she would be freer to enjoy the experience.

We come to wine with a template in the form of a collection of tasting terms, and these words can influence our experience of the wine itself. Once we have the word *Gewurztraminer* in our minds, it's hard not to let this affect our perception of the wine in the glass. This is particularly dangerous for wine experts like me, because we have template descriptions for different wine styles that we all too readily rush to when we know (or think we know) what sort of wine we are tasting.

Indeed, some researchers think we move too quickly to words, rather than dwelling in the sensory experience. Dr. Melanie McBride, who studies intersensory learning involving smell and taste at York University, Toronto, notes: "In cultures where smell is a primary learning and considered a critical dimension of experience, they have more language/words for it because it is a higher priority for them. In the West, where developmental psychology theories such as Piaget's ages and stages still dominate our thinking, we see physical knowledge as a low stage we're to grow out of and grow 'into' social knowledge." In other words, in our culture we downplay the sensory experience itself and jump straight to words. This top-down cognitive application of previously acquired knowledge can overshadow what we are actually experiencing.

For wine tasting, the implications are obvious. If we jump straight to winespeak, we lose the sensory experience of tasting and smelling the wine. McBride gives the example of strawberries. "We had to stop smearing the strawberry on our face and mashing it into our mouths because this was considered a lower and more infantile way of relating with knowledge than using words, signs, and symbols. And so when the sensory stage is pathologized as being infantile, we basically stop a process of learning that should have continued." Of course, the intent in those who would have us eat politely and carefully is not to deny sensory experience, but the result is the issue: that unintentionally, we are encouraged to move on from the immediacy of the sensory experience itself.

What does this mean for teaching people about wine? "I believe we could radically redefine wine learning with far more of an emphasis on physical knowledge and tasting than language or theory," says McBride. "I think the real problem with wine training is that it's far too focused on the theory and rote memorization of terms and sticking with the 'grid' than it is with actually learning how to use your tongue, mouth, nose, and eyes to understand what's in the glass and in your body."

So next time you try a wine, pause before you allow yourself to rush to write a tasting note. Don't analyze the wine. Stop those words forming in your mind. Take a step back and dwell, uncritically, in the actual experience of the tastes, textures, smells, and flavors. Give the wine time to speak to you. Then, and only then, reach for your words. It will be a different experience, and you may well enjoy the wine more.

Mouthfeel matters

It's in the mouth that we really get to
understand a wine. Texture, mouthfeel,
elegance, finesse—they're all underrated.

WE SO OFTEN fail to understand wine because we are
looking in the wrong place. In a way, we are like the
drunk looking for their keys under the streetlight. A
passerby sees their evident distress and offers to help.
"Where did you drop them?" they ask. "Over there" is
the reply. "Well, why are you looking here?" "Because
the light's better."

Our wine education teaches us to describe wine
in terms of the tastes and smells it elicits. I love the
Wine Aroma Wheel as much as the next person, and
I've enjoyed sniffing my way through the sometimes
accurate, sometimes perplexing smells of Le Nez
du Vin, but this obsession with such descriptors can

leave us hunting under the streetlight when we might actually find what we're looking for somewhere else.[1]

"For me, terroir is issued through shape and feel, not smells and flavors," says Nick Mills of the Central Otago (New Zealand) winery Rippon. "If you give me a Volnay, or a Pommard, or a Nuits-St-Georges, or a Vosne-Romanée, I am going to have a much better shot at telling you which is which if I think about structure, shape, feel, and form, than about cherries and plums and smells and flavors."

Wine becomes a lot more interesting when we bring mouthfeel and texture into the equation, rather

---

1. A turning point in how we talk about wine came in the 1970s and '80s, with work done by researchers in the enology department of the University of California, Davis. In 1976, Maynard Amerine and Edward Roessler published a manual titled Wines: Their Sensory Evaluation. Their goal was to replace vague and fanciful wine terms with a set of more precise words, shifting from the likes of masculine, naive, harmonious, and presumptuous to a standardized vocabulary that was more analytic in nature. The next major step in moving winespeak in a more scientifically rigorous direction came in 1984 with Ann Noble's now famous Wine Aroma Wheel. It consists of three concentric rings designating wine smells, beginning with broad olfactory categories in the center (such as woody, earthy, floral, herbaceous) and gradually getting more specific, with subcategories and then actual smells in the outer ring. This was an attempt to get past the obstacle that most tasters face when they first try to describe wine: recognizing smells and putting them into words. It's a very useful tool, but it alone isn't going to help you write good tasting notes.

than focusing only on smells and tastes. Combine both approaches, and we have a much better chance of writing tasting notes that actually communicate something useful. We begin looking at the wine as a whole.

12

Just because I'm learning doesn't mean I'm stupid

Most introduction-to-wine books are deeply unsatisfying. In their well-intentioned quest to educate readers about a complicated topic, they frequently fall short.

IT'S GREAT THAT there are so many books aiming to introduce people to wine. But they are often unsuccessful. Why? It's because when people try to take on the role of educator, they forget that many of their readers are actually quite smart—and may in fact be significantly smarter than the educators themselves.[1]

Educators sometimes patronize their readers, making the mistake of thinking that because they are

---

1. It is hard for us to recognize that some people are smarter than us, especially if we are highly educated and have letters after our name. But it's true. Some people are astonishingly smart. Mercifully, most of the time we aren't smart enough to realize the excess smartness of such people. It would be hard to take.

in possession of a body of knowledge that their readers lack, they are smarter than their readers.

There are lots of things I'd like to learn more about. Cheese, for a start. And maybe coffee. And Japanese gastronomy. But I don't want people to tell me about these things as if I were a ten-year-old child. I can deal with the complex bits and the interesting stuff, but what I need is for the teacher to make these subjects accessible.

Accessibility doesn't equal dumbed down. We can have rich content, but it needs to be packaged in the right way. It needs to avoid reliance on a corpus of background knowledge. And it needs to be interesting.

Rather than *Cheese for Dummies*,[2] I want to read *Cheese for Smart People*. Smart people who just don't know much about cheese yet. And one day I'd love to write an introduction to a wine book that doesn't talk down to my readers, but recognizes that they are smart and that they want to read something interesting.

---

2. I should note here that I'm not taking a pop at *Wine for Dummies*, which, despite the title, is actually a good, solid introduction-to-wine book.

13

Wine as an aesthetic system

Wine appreciation doesn't exist in isolation, but is part of a wider aesthetic system. We decide together what is great about wine, through our interactions, our discussions, and our learning.

HOW DO WE decide which the best wines are? Is it merely a personal, subjective decision, with no applicability to others? On one level, this sounds fabulously egalitarian. You are the consumer: you get to decide what is right for you. If you like a wine, who is to tell you that you're wrong?

Of course, I can't argue with this position. You really are free to like what you like. But if that's all there is to it, we may as well just go home. We've hit a roadblock. If I, as a critic, say that wine X is not very good, that it has some brett[1] that overwhelms

---

1. Brettanomyces, or brett for short, is a rogue yeast that contributes a suite of aromas and flavors to wine, typically described as animal, phenolic, medicinal, and spicy. At low levels it can add complexity, but usually its presence is considered to be a fault.

the fruit, and I award it 82 points, am I insulting the consumer who actually enjoys this wine? According to the egalitarian view, who cares what I think? We each make our own decision, and it doesn't matter what others believe: my view on the wine is merely personal—it's just autobiography.

Yet if we have wine critics sharing their views, they must believe that, to an extent, their views are normative. If as a critic you sell access to your tasting notes, your behavior indicates that these notes are not simply subjective and personal to you. Indeed, wine tasting isn't just subjective: there is a level of objectivity. But there's room in this objectivity for personal stylistic preferences. What we are dealing with in wine, in short, is an aesthetic system. As a wine trade, we sit down and talk about our experiences as we taste wines together, and through this process some wines emerge that are considered to be better examples of their type. With time, we come up with a loose hierarchy of producers that are considered to be better than others. And we continue to taste and revise our classifications and opinions, especially given that each year we have a new vintage to talk about.

And then there's the money. Some wines fetch higher prices, which would seem to indicate a certain level of objectivity to the taste of wine. Indeed, the 1855 classification of Bordeaux[2] was established on this basis of cost.

We acknowledge that there are individual differences in perception. Some people, for example, may be more sensitive to specific tastes, or to the smell of certain compounds. But we talk and learn together, and in the process, to some extent, individual differences are offset. We operate within an aesthetic system where decisions are made by a community of judgment—though to be sure, some opinions are allotted more validity than others. To reach a verdict, certain competencies need to be in place. We need to be tasting out of decent glasses, with the wine at the proper temperature. We need to be functioning well (be healthy, able to concentrate, free of distraction), and we need to have some experience with

2. This is a famous, enduring classification of the châteaux of the Médoc (also known as Bordeaux's left bank), which ranked the properties hierarchically, from first growths (premiers crus) through to fifth growths. Remarkably, even though the wine world has changed a lot since then, including a replanting of all the vineyards with grafted vines following the phylloxera crisis of the late nineteenth century, the classification is still relevant—and much admired outside of Bordeaux.

wine tasting. Then we can contribute usefully to the discussion.

Interestingly, the old fine wine aesthetic system of London wine merchants in the early to mid-twentieth century has come under challenge in recent decades. First, it has had to adapt to the emergence of fine wines from new world countries. Previously, these weren't considered to be part of the fine wine aesthetic, but gradually people have realized that by any sensible definition the best non-European wines belong in the same peer group as the established old world classics. Second, the emergence of the natural wine movement, as well as offshoots such as orange wines,[3] as brought an entirely novel sensibility to the fine wine aesthetic. This has caused some heated and entertaining debate, but natural wines are here to stay (however you choose to define them—another point of contention).

---

3. These are white wines that have been fermented on their skins, like red wines. It's a growing category. They often have more color than whites made the conventional way (where skins and juice are separated by pressing the grapes before fermentation), hence the term orange. They are also sometimes referred to as amber wines, or skin-contact whites.

We are still adapting to these various changes in the world of fine wines. But there is every sign that as a group, wine professionals worldwide will manage to forge a new wine aesthetic that is inclusive and useful.

14

Beauty is not the absence of flaws

It is only in the context of imperfection that we are able to discern true beauty. We should reexamine the way we think about wine faults.

THERE'S A JAPANESE concept called *wabi-sabi*. In simple terms, it is the idea that flaws, or features that are not beautiful in and of themselves, can bring out beauty, or that they are in fact *part* of beauty. The *wabi-sabi* aesthetic, which includes characteristics such as asymmetry, irregularity, simplicity, economy, modesty, and austerity, underscores the notion that beauty is transient, incomplete, and imperfect.

Compare this with the common Western aesthetic in which beauty is considered the attainment (or near attainment) of perfection.

An example of *wabi-sabi* in action can be found in a Japanese term that translates as "silence like a drop of water." Imagine you are sitting by a pool in a zen garden. It is a tranquil, silent setting—but only because the quiet background sound of dripping water brings out the silence, making it explicit. Without this subtle interruption, you might not have recognized and appreciated the silence.

In smell, consider the fragrance of jasmine, which contains an aromatic compound called indole. Alone, indole has a somewhat unpleasant smell, but in synergy with other compounds it helps create a wonderfully ethereal aroma. And then there is that celebrated perfume ingredient ambergris—the cetacean equivalent of a cat's fur ball. It is produced by sperm whales to protect the intestine from sharp items in their diet, such as squid beaks. These offending objects are coated in a substance that then congeals into a mass, which is then excreted through the whale's rectum. But there's no point in hunting sperm whales for ambergris. The transformation from intestinal debris into valuable perfume ingredient occurs in the aging process as it bobs around in the sea, sometimes for years. It is therefore only discovered when it washes ashore; if you find a decent-sized lump, you could be very well off indeed.

I'd always heard that ambergris smelled bad on its own, but as a component of perfume it created aromatic synergy and worked very well. But when I got to see some, and smell it, and put it on my skin, I found that far from smelling bad, ambergris is quite lovely, with a rich smell, much more bass than treble, with spice, vanilla, old wood, dried herbs, and maybe a whiff of smoke. It was also quite sweet. Three hours later, this tiny trace still smelled lovely, and it seemed that it wasn't fading. The experience was amazing. Now I understand why ambergris is so highly valued by perfumers: it isn't just a fixative, or a foil. It is something essential. Still, it remains an example of beauty coming from somewhere that otherwise would not be associated with beauty: the rectal excretion of a whale.

In a similar vein, let's consider music. What makes for beautiful music? Those engaging this question suggest that music follows mathematical order. But not perfectly: that would be boring. Music causes the release of the reward chemical dopamine in the brain, in a similar way to other pleasurable stimuli such as food, drugs, and sex. This dopamine release occurs both at peak emotional moments, such as a melodic hook or chorus, but also in anticipation of these good bits.

Good composers or songwriters tease us, showing us briefly the notes or chords that we want to hear, and then avoiding them until the main hook or chorus. They build up expectation, perhaps even diffusing it by including unexpected notes or chords. In a sense, they create something more beautiful by avoiding the obvious expression of what we would classify as beauty. It is the "flaws" or "deviations" that underscore the beauty when it is finally revealed. Or perhaps it is deeper than this: the flaws and deviations may well be part of this beauty.

Let's take this into the arena of flavor. Tasted alone, salt is not pleasant; it's aversive. But at the appropriate level it adds life to food, creating a tension or balance. The same goes for lemon juice, which alone is too acidic to be pleasurable, but which acts to create a needed balance in some dishes. Sweetness alone is cloying; counter it by adequate acidity, and it's lovely.

Facial attraction is a further example of the complex nature of beauty. Evidence suggests that the rules of facial beauty in humans are hard-wired, with such judgments being cross-culturally consistent among adults and children. Young babies, it has been noted, will spend longer looking at attractive faces than at less attractive ones. It is not an absolute property, of course, but there do seem to be guidelines governing what sort of faces people find desirable, and evolutionary explanations for why this might be so. Certainly, a beautiful face is something we find very compelling. Yet attempts by scientists to produce the "perfect" face result in a composite face that, while certainly striking and attractive, looks slightly bland and is less attractive than you'd expect it to be. It is the presence of subtle flaws that brings out real beauty. We love faces with a bit of character. Most of all, though, we are attracted to faces where we know the person behind the face.

How does this relate to wine? I think that the most attractive, compelling wines are those that contain characteristic elements which, if they were to be in a different wine context or present at higher levels, might be considered faulty. It all depends on the wine, the context, and the consumer. When is volatile acidity too high? When is greenness good and when is it bad? When are tannins too grippy and firm? When are earthy, spicy characters offputting? When is savoriness or gaminess too prominent? When is new oak too obtrusive? When does a wine pass from maturity to senescence? Many of these characters are present to a degree in some of the world's greatest wines.

My favorite "flaw" in wine is reduction, the term we use to describe volatile sulfur compounds in wine. These are produced by yeasts and undergo changes in the wine with time. At the right level, a bit of reduction can add real complexity. I especially like wines that have started out life quite reduced, and then this reduction has resolved and they carry subtle traces that add real interest: the "ghost" of reduction.

Perfect, flawless wines, technologically made, with no edges at all, tend to be soulless.

15

We drink wine for many reasons: flavor is just one of them

Often, we make the mistake of assuming that wine is all about the taste. The liquid in the bottle is only part of it.

I FREQUENTLY TASTE wine blind—that is, without knowing what wine I am drinking. This act is an interrogation of a liquid in a glass. And when the tasting is double blind (i.e., when you have no idea what wines are being served—at least in single blind tasting you have a list of the wines on offer, though you don't know the order in which they're served), the only cues available are the color of the wine, its smell, and its taste. It's quite hard to do well.

Tasting wine blind has some advantages. It frees you from any preconceptions you may have, and makes you focus on the actual properties of the liquid in the glass, rather than imposing some sort of prototype on the wine. For example, if you know the identity of the wine, you might use your knowledge of the wine to fill in gaps or even to override the signals your senses are delivering you. Sometimes this leads to nice surprises. Say you have decided that you don't like Pinotage, that uniquely South African grape created by crossing Cinsault and Pinot Noir in the 1920s.[1] Then you taste a wine blind, without being told it is Pinotage: it's an aromatic, lighter-style red with a lovely, supple personality, and you like it quite a bit. *It's almost Burgundian,* you say to yourself, *I wonder what it is?* Then the reveal. Suddenly you have to change your mind about Pinotage. It can make horrible wine, but now you know that it is also capable of making something delicious.

Knowledge changes the nature of the experience itself. Tasting sighted, for example, your knowledge

---

1. Pinotage, created by Professor Izak Perold in Stellenbosch, almost didn't make it. Perold planted the seeds from his experimental hybridization in his residence garden at the university in 1925. Two years later he left to work with the KWV, leaving the vines growing in his now untended garden. This new crossing was saved by a young lecturer, Dr. Charlie Niehaus, when Perold's garden was cleared out, and the four plants were replanted in the nursery at Elsenburg Agricultural College by Prof. C. J. Theron. Mean-spirited Pinotage haters might have preferred that Niehaus and Theron hadn't bothered.

of the producer might influence the way you interpret the sensory cues. A Pinot Noir from well-known
producer X, who you know is keen on whole-bunch
ferments, might reveal a strong green note lurking
under the fruit, which you interpret as coming from
stems and thus count in the wine's favor. If you were
to taste the wine blind, however, you might be less
receptive to this green character.

The disadvantage of tasting blind is that often
the best reading of a wine is one made when there is
knowledge of that wine: the producer's intent, the
region, the vintage characteristics, the track record
of the wine and so on. This information is useful
in interpreting the sensory information. The great
wine scientist Emile Peynaud once said, "Great wines
tasted blind often disappoint." Cynics might interpret
this as suggesting the emperor has no clothes, and
that a lot of great wines are living on their reputation.
I'd take a more positive stance and say that it's only
when we know what we are drinking that we can enjoy
a wine to its fullest. I might taste a really good wine
blind, but then when I know what it is I set myself
free to enjoy it in all its glory, confident that I am
drinking something that the community of judgment
in the wine trade has recognized as something special.

Because wine is more than just a liquid in a glass. Normal people don't drink wine independent of context. The chemical properties of a wine do matter, but everything around the wine also contributes to the experience of drinking it. Factors seemingly as trivial as the label, the bottle, and the capsule and cork matter a good deal. The drinker's expectation—perhaps spurred by some knowledge of the wine, producer, or region—matters too. The drinking occasion is hugely significant, as is the mood of the drinker.

We drink wine for a range of reasons, and not all to do with its flavor characteristics. And I do not restrict this only to the fine wine dimension, or the habits of the engaged consumer.

16

A virtuous intoxicant

*With its alcoholic content, wine is an
intoxicant. But we can make an argument
that it is, to some extent, different from
other such substances: it is a virtuous
intoxicant.*

WINE CONTAINS ALCOHOL, and alcohol gets people
drunk. Some people get very drunk and do bad things.
Others drink too much too often and develop health
problems. Society, generally, has a problem with
intoxication, and this is a problem for those who make
and sell alcohol.

In the UK, images of provincial town centers
turning into binge-drink-induced war zones on
Friday and Saturday nights have heralded the call for
more control of drinking. Young adults are turning up at hospital consultant appointments with
end-stage liver failure after an adolescence of boozy
excess. It's not at all pretty.

The shadow of prohibition looms over the drinks industry, as the legitimacy of a drug with such a bad social rap is called into question. Can we justify using for our pleasure a tool that in the wrong hands causes so much pain and trouble? Wouldn't the right thing be to relinquish our pleasure, in a utilitarian bid to alleviate the high social cost of alcohol abuse?

It's here where we make the tentative case that not all intoxicants are equal, and that wine is different. It's a slightly flimsy case, but there *is* a point to be made here.

Would you drink wine were it not for the alcohol? This is an odd question: in practice, removing alcohol from wine changes it in significant ways, and I've yet to find a reduced-alcohol or alcohol-free wine that I would enjoy drinking. But imagine that it were possible to remove alcohol from wine without affecting its flavor. Then, the answer is maybe, at least for me. It depends on the situation. Indeed, in some cases the alcohol is a burden. I really enjoy drinking wine, and I enjoy mild intoxication with friends and colleagues, but I don't enjoy being actually drunk. And other times I might want to drink wine yet stay sober—for example, if I am driving or I have an afternoon's work ahead of me. The point here is that we are interested in the properties of fine wines as much as (if not more than) we are in the intoxication they create.

It's for these reasons—the fact that wine is enjoyed as much for its sensory qualities as for its alcoholic content—that the late philosopher Roger Scruton described it as a "virtuous intoxicant."

A few years back, I heard Scruton talk on the philosophy of intoxication. He began with two questions. First, he asked, is intoxication a single phenomenon? In other words, is the intoxication induced by wine the same as or different from that induced by whiskey or cannabis? Then he asked whether intoxication is something that philosophers should even be exploring. If intoxication can be wholly explained in scientific terms, this leaves philosophers nothing to say.

While science can explain the physiology of the drunken state, Scruton argued, there is more to intoxication than just drunkenness. His idea was that the experience of drinking wine is intoxicating in itself, quite separately from the physiological effects of the alcohol it contains. So when we ask about intoxication, we are indeed asking a philosophical question. We can't make a direct causal connection between the state of intoxication and wine itself.

Scruton used the analogy of a football fan to illustrate the relationship between intoxication and wine. The excitement of the fan watching their team play is caused by the football match, but it isn't a definable physiological condition. "Intoxication induced by wine is directed at the wine in the same way that the excitement of a football match is directed toward the game," he said. But it is impossible to make a direct link between the game of football and the state of excitement in the fan.

Scruton then considered the relationships between our different senses. Thomas Aquinas famously distinguished the cognitive senses of sight and hearing from the "noncognitive" senses of taste and smell, a division that Scruton thought is still helpful today. He distinguished, on this basis, sensate and aesthetic pleasures: "The taste of wine is sensory," he said; "poetry is intellectual." Intoxication is considered sensate and not aesthetic. Along similar lines, according to Scruton, "a visual experience is a representation of reality, whereas taste and smell are not like that." This is reflected in the difference between cogent accounts of paintings and the imprecision of winespeak.

Then we are led to consider some more profound aspects of intoxication by wine. "Intoxicating drink is a symbol of and a means to achieve an inward transformation," Scruton said. "From ancient times wine has been allotted a sacred function. It enters the soul of the person drinking it."

To emphasize the special nature of wine, Scruton offered a four-way classification of stimulants:

1. Pleases us but doesn't alter the mind.
2. Alters the mind but gives no pleasure.
3. Alters the mind and pleases us.
4. Alters the mind by the act of pleasing us.

An example of the first is tobacco, which has some mental effects but doesn't actively alter the mind. The second is illustrated by drugs that we swallow purely for their effect, taking no pleasure in the drug ingestion process itself. Class 3 are stimulants that are mind-altering and that give pleasure in how we ingest them, such as cannabis or alcohol. Class 4 includes wine, where it is in the act of drinking that the mind is altered.

Alcohol in general and wine in particular have a unique social function—which is what the fourth class alludes to. Many of the social contexts we have devised are aimed at limiting consumption—and hence intoxication—by controlling the rate of intake. The buying of rounds of drinks in the pub and the circulation of wine at a dinner party are examples of this.

"The qualities that interest us in the wine reflect the social order of which we are a part," Scruton said. Wine is not simply a shot of alcohol. At its heart is the transformation of the grape in fermentation, followed by the transformation of the soul under its influence. The Greeks described fermentation as a "work of God," and this notion is reinforced by the fact that humans bring to this process the skill of husbandry: we aren't actually "making" the wine; we are creating the conditions for it to make itself.

Truth is an important component of wine, its effect present and revealed in the flavor. Thus wine has a quintessential honesty. There is "truth in wine," but this is truth for others, and not for us: as we drink wine, each of us reveals more of ourself to others; we talk more, and more openly. Wine is quite unlike other mind-altering drugs, which are dishonest in nature, because they claim to elevate the perception of the user such that the user enters a transcendental realm. These drugs lie in that they tell us about another world outside our own. Instead, wine tells us about the true world, the one we live in, revealing more about it.

I like the idea of wine as a "virtuous intoxicant." According to Scruton's view, there is something special about wine: it isn't like other drinks. As an indirect support of this idea, the consistent and long-standing role of wine in culture and religion does suggest that it is a unique substance.

17

There is always another wine

*Supply-and-demand imbalances mean that
every now and again, old favorite wines
are no longer affordable. Still, there are lots
of new ones to discover. Friends move on;
you make new friends.*

MY FIRST CASE purchase of wine was back in 2000.
I'd gone to a Rhône *en primeur* tasting at UK wine
merchant Bibendum in London, focusing on the 1999
vintage. At the time, as a relative novice, I was obsessed
with northern Rhône Syrah: I just loved the combination of floral cherry and blackberry fruit with meat,
spice, blood, and iodine, together with some pepper.
(I still do.) Here was a region whose imprint on the
wine flavor was unmistakable. I tasted one wine that
evening that blew me away: it was a cask sample of the
1999 Jamet Côte Rôtie, which was selling at £240
per case. For me, that was a lot of money, so I split
a case of this and a case of the cheaper Domaine du

Colombier Crozes-Hermitage with my friend Nick Alabaster. I now wish I had bought more. The wine I got for £20 a bottle now sells for around twenty times that amount. (It helps that 1999 was the best vintage of our generation, so I take comfort from knowing that I was right in my initial assessment of this wine.)

Since those days, the northern Rhône has become a lot more popular, and Jamet one of the superstar producers. Supply of Côte Rôtie, moreover, is limited—the appellation has just over 200 hectares of vines. So it makes sense that prices have risen over time.

This story has been repeated across many wine regions. Famous wines, such as the top names in Burgundy and Bordeaux, have seen prices rise steeply. Wines that were once in reach of those on professional salaries are now the preserve of the wealthy. It's a two-pronged problem: not only are prices going up, but as certain producers suddenly become famous, their wines become harder to get hold of—they simply disappear before you can get anywhere near them. Think of previously affordable and reasonably available producers such as Gonon in the Rhône, Overnoy in the Jura, Roulot in Burgundy, and Clos Rougeard and Richard Leroy in the Loire.

For those who have been buying certain wines for many years, it's tough when you recognize you can no longer afford them. It's also a bit problematic when you pull a wine from your cellar for dinner and realize that the market value of that bottle is equivalent to a second-hand car or a holiday in the sun. This is a sad thing, but we need to be positive: there is always another wine, and the discovery of new favorites is enjoyable, and keeps us from getting stuck in a rut. As for those now-expensive wines that you once bought cheaply: drink them with friends, and forget about their market value.

This is one of the things that I love about wine. It is dynamic. Even those bottles you have stored away are always changing, and have a finite life. Old producers disappear, go bad, or become inaccessible. New producers emerge. While there are perhaps limits to exceptional terroirs—and as we have discussed, you need a great terroir to make a great wine, even though great terroirs don't guarantee great wine—we haven't yet explored enough to reach the limit of privileged sites for winegrowing.

Occasionally I will stretch my budget to try one of the acknowledged great wines. Splitting the cost with friends is a good way to do this. After all, it's important to benchmark, in order to provide context. More commonly, though, my wine buying is focused on areas and producers making great wines but not at crazy prices. Fortunately, there remain many of these. In the northern Rhône, Cornas used to be thought of as rustic and not very good, but now those of us priced out of Côte-Rôtie have discovered a number of growers making excellent wines there. Yes, Thierry Allemand is now expensive, but there are quite a few growers whose wines are still in spending range. Then there are the producers who have been finding good sites in Crozes-Hermitage and Saint-Joseph. Aside from well-known areas in France, there are any number of great terroirs waiting for someone to show them some love, understanding, and allegiance. Think of Marcillac in the southwest, or the good terroirs in the south of Beaujolais, or the Auvergne with its volcanic soils. We have only really begun scratching the surface. Then there's Italy, with its mysterious array of soils and grape varieties. There's more than a lifetime's exploration to be had there. Let's not mourn for what is lost, but continue in our journey with open minds, a curious spirit, and a sense of gratitude.

18

Tasting transforms us

Wine: that beautiful liquid, rich in culture and interest, with its own transformative powers that set us free to enjoy it, interrogate it, and explore its many dimensions.

THE ABOVE IS a quote from my book *I Taste Red*, which explores the perception of wine. Sometimes, when we drink wine, we forget that it contains alcohol, and the point that I was making here is that wine alters us as we perceive it.

Our relationship with food and drink is intimate. We consume them, and at least a part of what we consume becomes part of us. In most societies, moreover, eating and drinking are sociable activities. We do them together, and there is a degree of intimacy about the endeavor. We might share with those around us our impressions of the wine we drink, as we consider its sensory properties: What does it taste like? What does it remind us of? What expectations do we have of the wine, knowing where it came from? And are those expectations met?

In the process of enjoying the wine, the alcohol slowly works its way into our bloodstream and begins its transformational magic. We start noticing a mellowing effect, as if the wine is unlocking parts of our mind. We are set free by this social lubricant to be more fully present in the moment, and some of our inhibitions begin to be shed. This, in turn, changes the way we perceive the wine. We are on a journey.

Consumed with friends, with a degree of moderation,[1] wine is a gift. It is healthful and fun. Its transformative effects help us to relax into each other's company, and if we are paying attention to the wine, assist us in its interrogation. Perhaps this is why people talk so much about a wine opening up as it is consumed. Yes, the wine may be changing as it is

---

1. Moderation is hard to define here. A lot depends on the setting, and the individual concerned. I have had many joyous evenings where we have each consumed far in excess of a bottle, and where at the end we could be described as somewhat drunk. But in those situations, among friends, I rarely recall anyone drinking to the point where they were out of control. For some people, a few glasses can be excess, and for others it is better not to drink at all.

exposed to air after the bottle is opened. But maybe what people are noticing is the way the wine is changing them, and how this affects their perception of the wine. With one sip of wine, we see a certain facet of it; with another sip, we gain further insight. It's like when you set up the fingertip touch detector on your phone or laptop: you have to touch it repeatedly, each time at a slightly different angle, until the device gets a full picture of your print. As we drink, we begin to ask different questions of the wine, and in many cases these questions act as fresh lenses, revealing new facets or casts of the wine. Take the wine into your body, and you are transformed.

19

We are not programmed to like certain wines

We differ in our preferences for food and drink. We also differ in our biology. But while this means we likely experience some wines differently from others, it's simplistic to suggest we are programmed to like certain wines.

WE HAVE A problem in the wine world. There are just too many wines. Customers entering a shop are faced with a wall of wines. In restaurants, they are presented with a long wine list. How are they to choose?

Data is our friend.[1] It is used to connect customers to products. Every day, for example, I encounter

---

1. As a science editor for fifteen years, I always insisted on the correct use of the word data. It is the plural of datum, so the correct way to say this is "data are" rather than "data is." But so widespread has become the misuse of data, treating it as singular, that now it is accepted as the norm—and since this is the way most people use it, I'm joining in. Even though it rankles a bit. It's amazing how attached we become to grammatical rules.

targeted advertising that employs data gleaned from my web browsing and social media use to try to match me with products I might be interested in.

Why not use data to match consumers with the wine that is perfect for them? What about segmenting the population along lines of preference? It sounds like a smart idea. The starting point here is biology. People are different in many ways, including their flavor preferences. Ask a roomful of people if there are any foods they don't like, and any that they especially do like, and you'll probably find that most people can list half a dozen items in each category, mostly different.

Can we find a biological basis for these flavor preferences? If so, then perhaps there's a way to segment wine drinkers, and pair them up with wines that they will enjoy.

The idea that we live in different taste worlds dates back to 1931, when A. L. Fox, working in his laboratory at DuPont, made a momentous discovery. He'd been synthesizing a novel chemical, and knocked over a vial full of it. As the dust flew into the air, his colleague remarked on how intensely bitter it tasted. Fox, however, couldn't taste it at all. Excited by this discovery, Fox ran around the building testing people's reaction. Sixty percent could discern the bitterness, while the remainder found the substance tasteless. The chemical was phenylthiocarbamate (PTC). Together with its close relative propylthioura-cil (PROP—which is a bit safer because it has fewer potential side-effects), PTC separates the popula-tion into groups of tasters and non-tasters. In the 1990s, researcher Linda Bartoshuk coined the term *supertaster* to describe a subset of tasters who have heightened sensitivity not only to PROP/PTC, but to a broad range of different flavors, such as the burn of alcohol and the bitterness of coffee. Bartoshuk describes them as living in a "neon" taste world. While it was found that the different sensitivity to PROP is genetic, and depends on a variant of the TAS2R38 gene (which codes for a taste receptor), there also seemed to be a correlation between greater taste sensitivity and having more taste buds on the tongue. Could it be that supertasters make better wine tasters?

Since then, new research has shown the PROP account to be a little simplistic. Humans have twenty-five bitter taste receptors, of which TAS2R38, the one that detects PROP, is just one. TAS2R31, for example, detects the bitterness of artificial sweeten-ers acesulfame-K (AceK) and saccharin; TAS2R31 detects the bitterness of quinine, and also corresponds to liking grapefruit; and TAS2R3/4/5 can explain the bitterness that some people experience with alcohol.

Then there are specific anosmias—odors that some people just can't smell—that are relevant to wine. The OR10G4 gene explains differences in responses to guaiacol, a smoky aroma that is found in wines suffering from smoke taint, and whatever version of this gene you happen to have will predict how intensely you experience guaiacol.[2] Variants in the OR2J3 gene influence how intensely people experience cis-3-hexanol, a grassy aroma. Variants in OR5A1 affect the way the floral-smelling beta-ionone is experienced—with a genetic association that means that roughly a quarter of the population are insensitive to the violet/floral aromas found in some wines, and will probably like these wines less. And there's the common smell blindness to rotundone, the peppery character found especially in Syrah, which a fifth of people don't get.

---

2. Researcher John Hayes, who specializes in these individual differences in perception, thinks the sensitivity to guaiacol may also influence sensitivity to other compounds associated with the spoilage yeast Brettanomyces, including 4-ethylphenol and 4-ethylguaiacol.

Yet more variation in the population involves salivary flow rate (affecting how intensely the dryness from tannins is experienced) and enzymes called glycosidases (which unlock certain flavors in wine once the wine is in our mouths).

Some have claimed that all this variation means that we are programmed to like certain wines, and that therefore that it's possible to segment the population and target individuals with specific wines. I disagree.

Our preferences are extremely malleable, for good evolutionary reasons. Innate preferences for nutritious, high-calorie foods are pretty universal. But we possess the ability to acquire novel tastes. Our sense of flavor combines with our memory, allowing us to explore novel, possibly nutritious food items, which then become new flavor preferences. (The memory bit is important, because we need to reject quickly things that have made us sick in the past, an inevitable risk of exploring new food sources.)

Certainly, this has been my experience. Today I like strong cheese, but fifteen years ago I wouldn't eat cheese at all—I just didn't like the taste and texture. The cheeses are the same as they were fifteen years ago; it's my response that has changed. When I started drinking coffee as a teenager, I had it with two sugars and milk. Now if you put sugar in my coffee, I find it unpalatable. And with wine, I began with richer, sweetly fruited reds that were easy to understand, and only over time moved to more savory and complex wines. Now I wouldn't find the wines I loved at first all that interesting, and I wouldn't want to drink them. This sort of journey is not unusual. Who liked wine the first time they tried it? When it comes to flavor, many of the things I love now were tastes that I found challenging when I first experienced them. Experience has largely trumped biology.

But more compelling than the biological argument is that of variety and mood. The notion of testing my biology and then presenting me with the perfect wine "for me" is as daft as suggesting you can match me with my dream meal. Much of my wine drinking is mood driven, and much of it is situational. I drink a wide variety of wines covering a whole spectrum of flavors, just as I eat wide range of foods. Much as I love the idea of data helping people find the wines they like, it is just too simplistic, and ignores the complex factors involved in our choices of what we eat and drink. We aren't programmed to like certain wines.

20

Scores can be useful, but mostly they are stupid

The wine world changed when critics started scoring wine. Suddenly, it was easy for all to see immediately which the "best" ones were. Wine became competitive. But scores are a bit silly, and have gotten progressively sillier as grade inflation has set in.

AMERICAN WINE CRITIC Robert Parker changed fine wine forever when he began scoring wines on a point scale of 100 in his publication *The Wine Advocate,* which took off in the early 1980s. Suddenly, wine was democratized. Cash-rich yet time-poor folk with an interest in wine could suddenly see which bottles were worth checking out, without the need for bothersome learning about wine. Such was the success of Parker's approach that just about every critic now uses the 100-point scale, with an awful lot of critics vying to replace the recently retired king.

The popularity of wine scores suggests that they have some merit. People do find them useful. Immediately, it is possible to compare one wine with another. Tasting notes, too, have some use, but alone they fail to indicate exactly how thrilling the taster found the wines. Language is imprecise, after all, especially when it comes to describing flavor.

In society generally, we are used to things being ranked and scored. Movies, restaurants, hotels, and running shoes all now come with a rating, usually generated by means of customer reviews. If we want to buy something, we want to buy the best our budget allows, and the quickest way of making this outcome a likelihood is to see what gets the highest rating.

So it is not surprising that ratings for wine have taken off as they have. It's likely that had Parker not started the 100-point scale, someone else would have. Society wants ranks. And scores. The best!

I admit it: I score wines. I do it because everyone else does, and I want my readers to see quickly how well I liked the wine. But I am deeply uneasy with this practice; it makes me feel like I am somehow compromising my values, being a bad person. For that reason, I have great admiration for my young colleague and friend Christina Rasmussen, who, embarking on a wine-writing career that will no doubt take her far, has vowed never to score wines.

Even though I score wines, there is something about the practice that just seems wrong. It is an attempt to make something diffuse and indefinable into something focused and precise. The scale itself, 100 points, brings with it an impression of precision, a false sense of authority. Ninety-six out of 100 seems like a score that has been arrived at by some scientific, analytical process, not plucked out of the air after detailed appraisal of the wine.[1] To attach a score to a wine seems as silly as rating your friends or giving a number to a summer's afternoon spent walking in the countryside. It's possible to do so, but it seems weird.

Andrew Jefford, one of the wine-writing greats, expresses the dilemma well: "Scores for wines are philosophically untenable, aesthetically noxious—but

---

1. Related to this is the idea that if you are going to lie, lie precisely. Alcohol levels in wine are a great example. Most countries allow some latitude on declared alcohol levels. If you are deliberately understating the alcohol level on a label, do it precisely. If your actual alcohol level is 14.1 percent, and you declare it lower, then use two decimal points. If you write 13.27 percent, people will think you are giving the exact level, not a rounded-down number. (The reason for rounding down is that these days, people prefer lower-alcohol wines in general.)

have great practical value. Wine scores will, therefore, be with us for as long as human beings drink wine."[2]

So let's address the 100-point score, and its evolution. When Parker started, he used a much wider scoring range than current critics do. Gradually, score inflation set in, and with it score compression at the top end. We live in a growth culture, where everything always has to get better. (This is seen in the UK's school exam system, where the questions have gotten ever easier and the grades ever higher, such that the top grades no longer identify the brightest kids—and so a star system appended to the A has been introduced.) As other wine critics began to issue scores on the same scale, a sort of critical arms race ensued. Retailers and wineries, faced with a range of scores, naturally quoted the higher ones. This created a subtle pressure to score more generously. Eager for exposure, critics upped their points. Whereas 90 was once a good score, 95 has become the new 90. Scores for decent wines are now hideously compressed in a narrow band of 90–100. Awarding 100 points became

2. Andrew Jefford, "Jefford on Monday: The 104-Point Second Wine," *Decanter,* August 10, 2015 (www.decanter.com/wine-news/opinion/jefford-on-monday/jefford-on-monday-the-104-point-second-wine-15523).

a sort of marketing exercise for the critic: their maximum score becomes a talking point and is used by the winery in its communications.

Score inflation is a particular problem in Australia, of all places. I normally think of Australians as sensible, but some of their leading critics have gone a bit barmy, issuing absurdly high scores for even quite average wines.

Where do we go from here? Greater restraint is needed—perhaps a recalibration of some sort—if scoring wines isn't to become even more of a silliness than it already is.

21

Wine, be yourself

*There's nothing wrong with commercial
wines. The world needs good cheap wine.
But cheap wine doesn't have to try to mimic
more serious wine through trickery. Honest
wines are better than "better" wines.*

WHEN I STARTED drinking wine as a student, I was bottom feeding—buying cheap wines from supermarkets.
Most tasted bad. Today wine tastes better, even at the
bottom end. That's a good thing. But let's explore the
issue a bit further. I'd argue that generally speaking, we
don't want our wines to taste nicer, but truer.

A quick taste—a mouthful—isn't a very meaningful way to assess a wine unless you have a lot of wine
expertise. Novice or low-involvement consumers
don't taste wine the same way experts do. They focus
on what is in the glass, and on the sensory clues that
are present. Experts are able to marshal their knowledge and use their cognitive abilities to help them

make sense of their perceptions. Normal people find tasting lots of wines at once bewildering because they don't have such mechanisms to help them parse what they are tasting.

When such consumers are presented with a range of wines and asked to state a preference—which tastes nicest?—it's not surprising that they choose strongly flavored yet gentle wines. For reds, this means sweetly fruited and deeply colored wines with low tannin and acid, and even some actual added sugar. They taste "nicer."

But let's make a comparison with cheese. I don't want someone to take my cheese and make it taste nicer. I want Comte that really tastes of Comte; I want Cheddar with a strong spicy tang; I want goat cheese that's a bit alarming at first.

If you give twenty average cheese consumers something that experts would consider to be a range of great cheeses, and then slipped in a mild, creamy cheddar made from pasteurized milk and with no real flavor, they might well prefer that. But where does that leave us? Does it mean that in order to sell cheese we should strip it of some of its flavor to make it more palatable? No, because we can quite easily live with the idea that there's mass market cheese sold cheaply in supermarkets, and there's the real stuff with proper flavor and provenance. The latter is what food writers are interested in. Consumers seem to be able to accept that the cheese market is segmented, and that the big bricks of cheddar with no flavor serve a purpose, while the expensive "proper" cheeses serve another, and if you are interested in flavor, you buy the latter.

With wine, it's largely the same. Ask a group of novice consumers to taste a range of wines, and as they sip their way through, they might well prefer the big-brand red with 10 g/l of residual sugar and no nasty tannins. Smooth, sweet, and tasty. Does that mean that there should be more of these wines on the market? Should we be making more off-dry reds because that's what a large segment of the market wants?

This isn't a straightforward question, because the answer may be different for different segments of the market.

Generally speaking, though, as with cheese, I don't want my wines to taste nicer: I want them to taste truer. Where there are wines of terroir—expressing a local flavor—I want to buy a wine that tastes of where it's from. That's what makes wine interesting. Of course, this local flavor is only partly derived from the site, which is the conventional understanding of terroir. But terroir as also an interpretive act: it's site *plus* the variety (or varieties) and the choices of the winegrower. Local cultural practices can contribute to the local flavor. Some places have more local flavor than others. That's just how it is. With wine, if you have a local flavor, no one can copy you. You are potentially able to rise above the mess that's the price-sensitive bottom end of the market.

Say you are a producer in Fitou. It's not easy to sell Fitou: there's a lot of it made, and it's quite cheap. What do you do? You can resort to winemaking trickery and add some post-ferment sugar, thus making a wine that tastes better (according to average consumers). You might be able to sell your wine to a supermarket more easily as a result, but will you get any more money? No. You'll just get to empty your tanks and, if you're lucky, not take a loss on the wine. The alternative is trickier to achieve but could lead to long-term success: make really good Fitou that tastes of the place, and transcend the appellation by supplementing the regional brand (Fitou) with your own brand. If you make excellent wine, and establish a good route to market, you have a chance of making money and escaping the dreadful bottom end of the market where producers merely survive (if they are lucky) but never succeed.

Make your wines taste truer. Not nicer.

22

The sadness of spoofulation

*There aren't all that many special places
to grow wine grapes. It's a tragedy when
a privileged terroir is used to make sweet,
oaky, international-styled red wines.*

THE TERM *SPOOFULATION* was launched on the internet somewhere, perhaps in the late 1990s. I'm a fan of
clever neologisms, and this is one of them. Although
I'm no lexicographer, I'll attempt a definition.

Spoofulation refers to the production of a spoofy
wine. A spoofy wine is one that has been processed
in such a way that it is no longer true to its origins,
but tastes made-up. The winemaking imprint has
obliterated genuine character through a combination of ripeness, sweetness, and perhaps also oak.
Other winemaking trickery may have been used with
the intention, no doubt, to make the wine have a
broader appeal, but to distinguished palates it tastes

horrible—confected, sweet, and placeless. A spoofy wine may be inexpensive, or it may be expensive. You can almost (but not quite) forgive spoofulation in a cheap mass-market wine (people have to make a living, and as someone once said, no one ever went broke underestimating the taste of the American public). But it is entirely unforgivable in an expensive wine. Especially expensive wine from a good terroir.

One of the tools of spoofiness is a grape juice concentrate and skin extract known as Mega Purple. I'd heard of this magical substance, and finally an honest winemaker showed me what it looks like. This is (supposedly) often used in California to add color to wines, as well as a little sweetness, and has caused quite a stir—perhaps, in part, because of the name. Regular grape juice concentrate is made by concentrating unfermented grape juice, typically by boiling it in a partial vacuum, which lowers the boiling point to a temperature where the flavors aren't completely cooked. The resulting superconcentrate is then added to wines that have already fermented to dryness at the blending stage. Many commercial red wines will end up with 4–5 g/l residual sugar, entirely from the addition of concentrate. Some are as high as 10 g/l, which gives them a bit of perceptible sweetness. The popular Apothic Red is all the way up at 16 g/l. A little bit of sugar rounds out the palate and adds to the perception of fruitiness. It also masks harsh tannins and covers over a bit of greenness. Concentrate is typically 68 °Brix (i.e., 68 grams of sugar in 100 ml, or 680 g/l sugar): it's viscous and intensely sweet.

Mega Purple adds a bit of sweetness, but is mainly used to add color, made as it is from a teinturier (red-fleshed) *Vitis vinifera* variety called Rubired. Mega Purple is produced by Constellation-owned Canandaigua. It's expensive, at around $125 a gallon, but very little is needed, a typical addition being 0.2 percent. There is also apparently a Mega Red. The wine trade doesn't really like people talking about these products.

We now live in the era of the sweet red wine. It's incredibly common to taste commercial reds that have added sweetness. But these are cheap wines, and my bigger issue is with spoofy wines made from good vineyard sources. This is a great sadness; after all, the world isn't awash with great vineyard sites, so we should cherish the ones we have, and they should be allowed to express themselves.

What of the new wave of beverages that aren't strictly wines, but that have flavor additions?

Supermarkets now frequently have a specific area devoted to fruit-flavored wine-based drinks that are packaged to look like wine, but usually with brighter labels and in clear glass bottles. The motivation here is to make wines more accessible (often they are sweet) to the younger drinker. Perhaps they do act as bridges to bring new consumers into the wine category, and this is a good thing if that's really the case. But there is an attendant danger in reducing wine to a flavored alcoholic beverage. If wine is no longer special or different, then any profitability left in the category will be stripped out by the big retailers who control the route to market.

23

Too many commercial palates

*The wine trade is chock full of talented
tasters, but too many have commercial
palates. They are skilled at differentiating
among commercial wines, even very good
wines, but can't differentiate top-quality
commercial wines from truly serious wines.*

IT'S FAIR TO say that DIY is not my strong point. I'm
not terrible at it, but it's not really one of the things I'd
say I have a talent for. Partly, this is because I'm just
a little approximate. There's a saying, "measure twice,
cut once," and it's true. I rarely measure as accurately
as I should, and often find myself having to cut twice
or drill a second hole. My tendency toward approxima-
tion is also evident in my toolkit. Life is much more
straightforward in DIY if you keep your tools in the
right place, returning them immediately after you've
used them. This is especially true with screwdrivers.
You buy a full set, but then one or two go missing, and
you find yourself trying to use a slightly wrong-sized

screwdriver. It works, but not very well, and it's frus-trating. Screwdrivers are specialists, not all-rounders. Always use the right one.

The same is true of wine tasters. In order to learn about wines, we try to taste as many as possible, across all sorts of different styles and price points and countries of origin. Eventually, though, we find we have talent for specific styles of wines. We learn to specialize.

It has been my experience that the wine trade, by necessity, is full of people who are very good at tasting commercial wines.[1] This is their sweet spot, largely because this is what they are being paid to do, whether they are winemakers, wine merchants, buyers, or journalists. Theirs is an important skill: they have to taste lots of young wines at varying price points and make important commercial decisions about their merits. In many cases they have to project six months or a year into the future to when the wine will be consumed, and have a good idea of how it will have developed by then.

---

1. I am aware that the term *commercial* here is probably meaningless. If a wine is sold at a profit, it is by definition commercial. Therefore, fine wines are often the most commercial because they make the most money.

But if you put them in front of a range of top-quality commercial wines, on the one hand, and truly fine wines on the other, they are frequently unable to differentiate. In contrast, I know many collectors and enthusiasts who pretty much only drink what we would describe as fine wine, and they are able to distinguish the really fine wines from the very good commercial ones quite easily. They are a certain size of screwdriver. They would be hopeless at tasting through, say, a range of forty Sauvignon Blanc tank samples, as a supermarket buyer must do frequently.

Yet if you suggest to a professional with a commercial palate that their best-selling super-premium wine isn't a serious/fine wine, they tend to get upset. There is often no recognition of a meaningful difference. This is a problem.

24

Monsters aren't serious

There is a place for monster, ripe, bad-ass wines. It's just that they aren't serious. But so often the people who make them want them to be taken seriously, which instantly makes them into joke wines.

THE DECADE OF the 1990s has a lot to answer for. This period—an era we are only now recovering from—saw the emergence of monster wines. It almost ruined fine wine.

I'm a tolerant person, and I believe anyone should be free to make the wine they want to make. Indeed, when I was getting into wine back in the mid-1990s, I had quite a thing for big wines. In some ways, the bigger the better: I wanted flavor in my glass, and lots of it. I liked the idea of red wines so dense you could stand a spoon in them. Deep in color, sweetly fruited, mouth-filling, tannic, and with an obvious impact.

In some ways, there's a parallel between winemaking and effects processing on a guitar. As a teenager, I fell in love with music, saving up my money to buy an electric guitar. I taught myself to play. I had a real love for the sound of distortion: there was something about a power chord, and although I couldn't afford a Marshall valve amp, I could afford a distortion pedal. Later, I saved up to buy a multi-effects unit, which in addition to distortion and overdrive gave me delay, chorus, flanger, phaser, compressor, and reverb, among other modulations of sound. This unit came with a range of presettings that had been programmed by the engineers working for the manufacturer. They were fantastical, with too much of this, too much of that, and sounded like caricatures of famous guitar sounds. Totally unusable. It was through the process of programming in my own settings that I began to realize that while effects are very useful, less is often more, and what sounds good in your bedroom can sound terrible at concert volume. Later still, I began to realize the merits of a more-or-less raw guitar sound, with the unaffected guitar plugged straight into a good guitar amp. With analogy to guitars and effects, too much ripeness and winemaking trickery results in essentially undrinkable wine.

So in my wine-drinking journey, I soon began to realize that there was more to wine than volume. I fell out of love with bigger wines, especially where higher alcohol and overripeness reared their ugly heads.

But what of Port, you say? A young vintage Port is very ripe and sweet, weighing in at 20 percent alcohol and around 100 grams per liter of residual sugar. Isn't there a bit of hypocrisy here? How can you bash a US-critic-lauded international-style red with 15 percent alcohol and lush sweet fruit when you've publicly spoken about how much you enjoy a good young vintage Port, with its volume knob well beyond the 11 that you criticize the international red for?

This is a question that demands an answer, and my answer comes in two segments. First, Port is a traditional wine with heritage and a track record for aging. I know if I cellar a good vintage Port, it will morph into something mellow and beautiful with thirty years in the bottle; and since it is still made now the way it was made in the past (with a few refinements, such as better-quality brandy spirit for fortification), I have no doubt that the modern wine will age just as well. I can't make that call to tradition for the new-wave monster reds, which I suspect won't age as well. And second, there's something about Port that makes it delicious and harmonious, with lots of sweetness and alcohol but also lots of tannin. On paper it should be big and messy, but it just seems to work.

Monster wines have their place, and some people enjoy them. I appreciate this, and I have no problem with it. The issue comes when people take terroirs capable of producing great wines and use them to make monster wines with no sense of place. This is a moral issue. There are relatively few places capable of making great wines with a strong somewhereness to them, so shouldn't they be used to make wines of place? And I take issue with people who try to claim that monster wines are serious, by giving them enormously high ratings and putting them in the same peer group as truly great wines. (We will discuss later on who gets to decide which wines are great.)

Let's use a sports analogy. You might be a handy golfer with a single-figure handicap gained at your rather benign local course. Anyone watching you would think you are quite talented: you strike the ball really well, and your swing has a nice economy to it. Then, by some bizarre twist of fate you find yourself teeing off at Augusta National, a course set up to discriminate among the very best professional golfers. Suddenly you are way out of your depth. A joke! You are utterly defeated by the course, and those fast greens make you look like a beginner. That's because there really is a difference between a talented amateur and a professional, but you'd need to know a lot about golf to spot that, before it was ruthlessly exposed on one of the world's trickiest courses. Those monster wines, whether they come from Napa, or the Barossa, or Mendoza, or Tuscany, may look the part and have price tags to match, but by claiming to be among the very best, they risk being ruthlessly exposed for what they truly are.

In the 1990s, some leading critics championed these big wines, and winegrowers realized that by picking later they could see incremental rises in their scores. By polishing their wines by élevage in new oak barrels, they'd get even more points. And rich wine lovers, drawn by the high scores, would pay a lot of money for these wines. The critics would also punish wineries for making wines that they did not consider ripe and seductive enough, or that were difficult in their youth. It was a perfect storm of wine stupidity, and the result was childish wines with limited aging potential. It pushed fine wine to the edge of the precipice. But the good news is that this collective insanity seems to be on the way out.

As with so many discussions, the debate about ripeness, big wines, and the influence of critics tends to polarize. We end up with trench warfare, filter bubbles, and confirmation bias. It's easy to forget that wine is, after all, just a drink. The reason I complain about the monster wines, however, is that wine is culturally rich and can be profound, and I don't want to see all this lost—and for a while, as the American critics became powerful, there was a risk that much that is significant about wine was under threat. So I'll stick with my position: monsters aren't serious.

25

The evil of overripeness

Overripeness in red wines is a grave sin that has to be covered up with acidification and oak. Often, sadly enough, it is avoidable.

SOMETIMES PEOPLE MAKE the mistake of thinking that because a little of something is good, more of the same will also be good. A great example of this error in thinking is with air conditioning. On hot, humid days of the sort where even a gentle walk will cause rivulets of sweat to flow freely down your brow and back, air conditioning is welcome. You experience relief when you walk into a restaurant and suddenly feel a downdraft of deliciously cold air. But I've lost count of the number of evenings when I've sat in a restaurant in Singapore, Texas, or Hong Kong and shivered under the icy blast of an air-conditioning unit turned up too high. Being cold is as bad as being hot. A little cooling is good; too much is unpleasant.

It's that the same with ripeness and wine. Back in the day, many European reds were made from grapes routinely picked underripe, with extra sugar frequently added to the must in order to boost the alcohol level. The challenge in the classical wine regions has traditionally been to harvest the grapes before the autumn rains, making picking decisions tricky. Often, the grapes were brought in before they were fully ready, picked on the basis of sugar ripeness, rather than allowing the flavors to develop fully. This was particularly true in Bordeaux and Burgundy. A good vintage was one where benign weather conditions allowed the grapes to be left on the vine as long as necessary for flavors to develop properly.

Along came the consultants and the critics, and the tendency to pick too early began to change. The critics began favoring wines made from grapes picked later, with riper fruit flavors and more volume in the mouth. The consultants began to advise properties to pick later, and to encourage maturity by reducing yields so the grapes ripened more quickly. Initially, this seemed like a welcome development, because it was correcting the problem of underripeness, encouraging those with ambition to take the risk of waiting.

But along came global warming and increased wine prices (leading to more competition and an economic reward for making better wines). The result? A perfect storm of ripeness. The pendulum swing that initially seemed a welcome correction kept moving, into the realm of excess. And to make things worse, Bordeaux consultants began to spread their lower-yield/later-pick mantra to the new world, where there had never been a problem with ripeness. Thus the term *physiological ripeness* (referred to cynically by some as "psychological ripeness") was born, and suddenly winegrowers were abandoning their refractometers and instead walking through their vineyards tasting the grapes, in the search of the right flavor.

Longevity, finesse, sense of place, and drinkability were traded for high scores and immediate gratification. These later-picked wines certainly had some deliciousness, especially if you were from a country with a taste for sweet. Have you ever had breakfast in America?

But they were problematic wines. Just as underripeness is a problem, so is overripeness, because it makes wines of style with flavor holes that need to be filled with acidification and new oak.

Is the pendulum swinging back? I think so, and it's a welcome return to the established fine wine benchmark of expression of place, balance, and the ability to develop positively with age (and not just survive).

I speak to lots of winegrowers and taste their wines with them. Seldom do I hear a winegrower say, "I wish I'd picked a little later." We've all gone down the cul-de-sac of overripeness without even realizing it, and it is taking a while to backtrack. While some wines seem destined to stay there forever, and some are forced there by circumstances beyond their control, for many, there is a simple way out: pick earlier.

26

Express the vintage

Vintage variation isn't a problem to be ironed out. By all means, combat the challenges of each vintage with gusto. But consider the vintage conditions in dealing with the wine in the cellar. Vintage variation adds interest when handled well.

WE HAVE ALREADY touched on the issue of vintage variation. Typically, when we talk about wine regions, we discuss climate and microclimate. But climate is an average covering many seasons. What a vine actually experiences is the changing weather of the year, which we describe using the term *vintage*.

One of the joys of wine appreciation lies in comparing different vintages of the same wine. You may have a favorite Bordeaux château—one from which you purchase a case each year. (This purchasing behavior, of course, represents a rather old-fashioned, traditional approach to wine, but I hope it illustrates my point. If you take exception, then please substitute your region of choice.) Initially, we can study vintage reports describing the viticultural year. Inevitably, the narrative here outlines certain challenges, but in the end everything turns out all right. In worst-case scenarios, victory is snatched from the jaws of defeat, by picking either before the rains, or between the rains—or, in a wretched year, by the mysteriously named Indian summer.

Then we have our first impressions of the wine, as a mere infant, drunk well before its time. Is it rich? Or lighter? Is it structured and tannic, or likely to be approachable early? Are the various elements in harmony, or do they fight each other, struggling for some in-glass dominance? We may decant the young wine to try to tease out more of its personality, or follow an open bottle for a few days to gain a sense of which direction age might take it.

From time to time we return to the wine, for we want to drink most of our bottles when we feel the wine is peaking. We make predictions, and hope that the wine keeps its side of the bargain. Each vintage is different, yet we maintain some recognition of the place that gave it birth.

The growing season itself is a dance between the weather and the winegrower. The weather makes a move, and the winegrower responds. It's always the weather that leads, though, and that is in control. Sometimes the winegrower is defeated, and the crop is reduced, or in the worst cases lost altogether. It is the possibility of triumph or disaster (or more normally, something in between) that makes wine-growing so compelling. It's like watching a game of sport, where victory is not guaranteed and there's a real chance of failure or setback of some sort, which makes the ultimate victory all the sweeter. But with wine, we can't tell whether it was a victory until much later—sometimes decades later. Until then, we can only follow with interest. Let us embrace the vintage, and be humble in its face.

27

Stay critical, but remember: there is room for everyone

If we are to act as critics, we need to express our opinions. But we also need to retain humility. We don't know everything; we can be wrong. We should be careful not to lose this awareness, and to maintain an inclusive attitude.

I'M QUITE OPINIONATED, as you can tell. In this book I've referred to wines as being evil or childish or spoofy.[1] But I hope that I retain self-awareness, and I know I can be wrong. And I am reasonably tolerant of other viewpoints. All of us need to maintain the mindset of the eternal learner and be ready to revise our

---

1. Spoofy is one of my favorite wine descriptors. I don't know who invented the term spoofulation, but it was popularized by the late Joe Dressner, a wine importer. A spoofy wine is one that is tricked up, over-ripe, confected, made to look more than it is through manipulation, and generally not the sort of thing I care for. This is discussed more in chapter 22.

attitudes in the face of new experiences and insights. Yet on this lifelong journey, we can't avoid expressing opinions just because we know there is more to be learned. Nor can we add numerous qualifiers to each of our opinions, which would rapidly become tedious. So we make our judgments, but recognize that there is room for alternative views; we celebrate inclusivity and are generally nice to those who disagree with us.

The emergence of the natural wine movement has created quite a bit of controversy, and a number of influential writers have been prompted to tell those who like natural wine that they are wrong. This is interesting to observe: I am perplexed by their behavior on a number of levels, most particularly why they care so much about putting others right, and also how confident they are that they are correct while those who like natural wine are in error. So I wrote a letter to them, which I published on my blog, and which I reproduce here:

Dear (insert name)

I have heard that you don't like natural wines.

I can understand that this must be quite distressing for you. But do not worry: I am a doctor (a plant PhD, not a medic, but who cares? This is the internet) and I am here to help.

You have spent a lot of time and money on your wine education. You have learned a great many objective facts about wine: its production, its history, its global spread, and how it is supposed to taste. You have a finely honed palate and can differentiate among poor, ordinary, good, and great wines. So I can understand how upsetting it is when some of your colleagues (who should know better) begin championing wines that fall outside this frame of reference. It just won't do.

The first thing I need to tell you is that you matter. You are one of the top wine authorities/emerging stars of the wine world/top restaurant critics (select as appropriate), and people are intensely interested in what you have to say. They want to know what you had for breakfast, your favorite sports teams, your taste in music, your preferred tailor, and recent novels that you approve of. And of course, which wine styles you consider to be legitimate or illegitimate.

Because you are truly important, people are especially interested in hearing about things that you don't like. It's different for me. There are quite a lot of things that I don't like. They include butter in my sandwiches, the Archers, the Rolling Stones, greed, Manchester United, Dermot O'Leary, UKIP, the *Daily Mail*, queuing, most Chilean Pinot Noir, and cheapness. But I'm not like you, and I don't think my readers really care terribly much about my dislikes.

Now that I have reassured you about your significance, the next step is that I need to encourage you to tell as many people as possible about your dislike of natural wine. It is important that someone of your stature should do everything they can to help stop the spread of this terrible movement.

The idea that people should be free to make up their own mind about which wines they prefer to drink is a dangerous delusion, and could lead to lots of people drinking bad wine and thinking that they enjoy it. You know all about wine faults, and from what I hear from others, pretty much every natural wine you have had has been faulty (by your definition). Unfortunately, most consumers haven't had the sort of wine education that you have, and there's a very real threat that they might not realize that, as they drink these wines, they are enjoying wines that are flawed.

The nightmare scenario? That people should bypass gatekeepers like you altogether, and begin to explore and enjoy wines without the sort of essential guidance that you offer. They will begin making their own minds up, and that could be disastrous. You have heard about the RAW and Real Wine fairs that have been held in London and elsewhere over the last few years. The rumor is that these fairs have been rammed with normal consumers who have had a great time drinking natural wines. I suspect (and hope) that this is just propaganda from the organizers, and that the few people who made it to them couldn't find anything even half drinkable.

So you need to keep telling your readers how bad natural wines are. Really scare them. Tell them that they are cloudy, feral, stinky concoctions, packed full of wine faults. Suggest that the people who make them are deluded hippies with long beards and no clue about wine. Liken them to rough farmhouse ciders.

I realize that this is a distressing time for you. There are people—smart people even—who like things that you don't. Please surround yourself with like-minded colleagues who share your insecurities about the rise of natural wine. Poke fun at the natural wine movement and its supporters at every opportunity. And remember: confirmation bias is your friend. You are smart, and some of your smart friends agree with you, so you must be right.

I have a tactic for you. If people complain about your negativity toward natural wine, then act as if you are the one being persecuted. Complain that others are insisting you should like these wines. How dare they suggest that your palate isn't sophisticated enough to enjoy them! Stop forcing these faulty wines on me! It is extremely unreasonable for others to suggest that if you don't like natural wine, then you should just leave others to enjoy it without pointing out how wrong they are. You can't stay silent!

Natural wine is just a fad. Give it a year and it will all have gone away, and things will be just as they were before—the nice cozy, compartmentalized, tidy wine world that you learned about in your studies.

28

A mystical transformation

Wine is made by microbes, but so often we forget about the importance of yeasts and bacteria in this mystical transformation.[1] That's an error on our part.

IF YEASTS AND bacteria were bigger—big enough that we could see them—we'd talk a lot more about them. Without them, even the most famous vineyards in the world would just be yielding grape juice. I've never tasted grape must from Romanée Conti,[2] but I'd wager that there's nothing terribly exciting about it. It would be very sweet, and taste like grape juice. But

---

1. Conflict of interest declaration: I occasionally do some work for Lallemand, a major manufacturer of selected yeasts, bacteria, and inactivated yeast products. The views expressed here are my own honest opinions, but it is only appropriate that as a reader you know of this connection.
2. For those unfamiliar with fine wine, this is Burgundy's most prized vineyard, making one of the world's most prized and expensive wines.

after a week or so of fermentation, and at some later stage, the second bacterial fermentation, and after a while in cask, we have one of the world's most compelling and sought-after wines.

Of course, the starting material—the grapes—is what makes the difference. But it's the role of the microbes to add some compounds of their own into the mix, and then we have wine in all its thrilling diversity and complexity.

Given the difference in taste between grape juice and the final wine, it's quite remarkable how stable site differences can be. Terroir is latent; it's the microbial activity that unfurls it and makes it a reality.

Let's consider the process in a bit more detail. Freshly harvested grapes have microbes associated with them: a range of yeast and bacteria. Some of these will have roles to play in the fermentation process. A wild fermentation—one where no cultured yeasts are added—is a dynamic affair. The first few days of fermentation are carried out by what are known as non-saccs. These are a range of wild yeast species that don't belong to the genus *Saccharomyces,* of which the species *S. cerevisiae* is the alcoholic yeast that finishes off almost all wine fermentations. These wild yeasts kick things off, but are eventually poisoned by the rising alcohol levels; after the wine gets to about 4 percent (sometimes a little more), *S. cerevisiae*—which was there all along but initially only at very low levels—then takes over and completes the fermentation.

The prevailing view used to be that *S. cerevisiae* is very rare in the vineyard and that most fermentations were finished off by winery-resident yeasts. But the picture is more complicated than that. Sometimes where cultured yeasts have been used in a winery, researchers have found that they are dominating wild ferments. But other studies have shown that *S. cerevisiae* strains specific to the vineyard are carrying out the fermentations.[3] It all depends, it seems.

And when it comes to *S. cerevisiae,* the strain does matter. Different yeast strains create wines that taste quite different, and for this reason there has been a lot of work isolating strains with desirable features,

---

3. There are many strong opinions on the role of wild yeasts in winemaking, and some people hold dearly to the "'truths" they were taught in winemaking school. The prevailing opinion used to be that ferments were almost always carried out by winery-resident strains of yeast, and that S. cerevisiae was rare to nonexistent in the vineyard. Once people have learned something, they tend to hold on to this knowledge and relinquish it only slowly.

culturing them, and then commercializing them in packets of active dried yeast. Winemakers must therefore decide whether or not to inoculate. Cultured yeast can be considered to be part of the toolkit that allows a winegrower to interpret their place in a particular way. In some cases, inoculating might result in a better expression of terroir. In others, allowing the yeast from the vineyard to handle the fermentation might be the best route to this goal.

For those winegrowers seeking a degree of control but who like the textural qualities and complexity of wild ferments, it is now possible to buy cultured non-saccs. These wild yeasts, such as *Torulospora* and *Pichia,* can be used in two ways. They can be inoculated sequentially, beginning with the wild yeast and following up with the desired strain of *S. cerevisiae,* or they can be co-inoculated in ratios that favor the wild yeast initially.

Let's not forget about bacteria. Almost all red wines require a second fermentation, as do many whites—for white winemaking, it is a stylistic tool, depending on the initial acidity of the wine and the grape variety. In most cases, malolactic fermentation is allowed to occur spontaneously, but you can also inoculate with specific bacterial strains. Because malolactic bacteria can have quite a sensory impact on wine, not always positive, many choose to inoculate for this fermentation.

While microbes make many positive contributions to wine, they can also have a negative effect. Bacterial action on ripe grapes in the vineyard and on the developing wine can produce volatile acidity (VA), which is acceptable and even desirable at low levels, but which can ruin the wine if levels get too high. Sorting grapes carefully and keeping caps wet during red wine fermentation can stop VA from becoming a problem, and as the wine matures, protecting it from oxygen also helps. Then there's the thorny issue of *Brettanomyces,* a rogue yeast that affects red wines far more commonly than whites. Here too, its savory, phenolic, animal-like aromas can be appealing at very low levels in the right context, but it spoils more wines than it improves. Keeping the pH low and adding sulfites at the appropriate times are a good defense, but it seems that some vineyard blocks are more prone to creating bretty wines. Poor cellar hygiene encourages it too.

Wine. It's all about the microbes.

Don't expect others
to pick up your tab

*The wine business—and especially
vineyards—must be sustainable.
You can't expect the next generation
to pick up your tab.*

WHEN I WAS at university, we used to go out a lot. This
was a campus university, and so we'd often head out to
the student union bar: it was cheap, and we knew a lot
of folks there. One of our regular crew was a guy whom
I shall call Bob. Now, none of us had a lot of disposable
income—we were students, after all—but we'd come
with enough money to tide us through the evening. All
except for Bob. He'd come out with a pound—enough
to buy one drink. What do you do? Well, he's a mate,
so you take turns paying for his drinks. Pulling off this
sort of stunt once or twice is OK: we all fall on hard

times. But do it consistently, and you're going to get noticed. It's unreasonable behavior. Don't expect others to pick up your tab.

I think this is a good guide in life generally, not just when we're drinking at bars. It's particularly true when it comes to the environment. Vines need a lot of spraying to protect them from disease and from insect pests. And they need nutrition. The cheapest way of doing this is to buy a bunch of chemical solutions and then spray. And to buy the strongest, most effective chemicals, so you can apply them less frequently. The mindset here is based on a firmly vine-centric view: this is the crop plant, and the vineyard is merely the medium in which the vine is grown. For a long while, viticulturists idealized the tidy vineyard, with neat rows of vines growing out of weed-free, bare soil.

But now we have a new view of the vineyard, seeing it instead as an agro-ecosystem, in which the vine is but one living entity. The soils, too, are alive, contributing to a healthy, balanced ecosystem that is self-correcting and less vulnerable to pests and diseases, and that produces grapes that make more interesting wines.

Most significantly, the chemically[1] farmed vineyard is not sustainable. And this is the deal breaker. Just as you can't go to a bar and expect your mates to pay for you each time, neither can we expect the next generation to pick up our tab. If we are custodians of a piece of land, how can we hand it on to our children in a worse state? That's what we have been doing with this planet, and it has to stop.

At the moment, truly sustainable viticulture—one where farming vines doesn't diminish the quality of soils and water—is seen as a nice luxury, something that's desirable if we can afford it. Or it is used as a marketing tool. Instead, we need to view it as the bare minimum. Once we are farming sustainably, we can start thinking about how to do even more, in order to hand the land down in a better state than when we took it on—and in the process, make better wine through better farming.

---

1. I acknowledge here there is a problem with this use of the term chemical, because organic and biodynamic vineyards are treated with copper and sulfur, which, though traditional, are still also "chemicals."

30

True to origins

*If you put the name of a place on a wine
label, the wine should taste of that place.*

TRADITIONALLY, IN THE classic wine-producing
countries—most notably France—wines were labeled
by place. (Mostly, they still are.) This is something
we should celebrate, because it recognizes one of the
great facets of wine: that is, wine can capture place. It
can taste of somewhere. But that raises the problem
of complexity. If you are a local, even if you aren't a
wine geek, you'll probably know the names of com-
munes and towns in your own wine region. If you are a
consumer shopping in another country, you'll need to
learn quite a few names if you are going to navigate the
wine selection with any degree of certainty. That's why
varietal labeling has caught on, particularly in the new

world. Napa Valley winegrower and tireless marketing powerhouse Robert Mondavi started the trend back in the 1960s: instead of using ripped-off generic terms (Chablis, Claret, Champagne, and Burgundy were routinely appropriated by Californian vintners at the time), he decided to put the grape variety on the label. To the casual consumer, this provides a brilliant cue as to what the wine might actually taste like. Instead of learning a gazillion place-names, all you need to do is become familiar with perhaps half a dozen grape names. I'm all for this sort of simplification.

But I'm a wine geek. So I generally like to see the name of a place on a wine label—with a caveat: if you are so labeling a wine, the wine should taste of that place. I'm all for atypical, brave, and experimental wines, but if you put the name of a place on your label, you are making a sort of unofficial contract with the consumer, in that you are giving them an idea of what the wine should taste like through the geographical location.

This goes beyond truth in labeling. If the taste isn't typical of a place, I feel it's more honest to declassify your wine, like Chianti Classico producers do. There, even if a wine comes from the Classico area, it's normal to have a range that, as well as those labeled Chianti Classico, has one that is labeled as IGT Toscana, if the wine deviates in flavor, for example by being made of Merlot aged in small new oak. There's a truthfulness to doing so, even if that wasn't the original intention when people started making Super Tuscan wines.

31

No new clothes

If you hate overripeness and obvious new oak (as you should), take care lest you end up praising a wine for the mere absence of these faults. It happens.

I'VE BEEN JUDGING wine for quite a long time. Judging wine blind is one of the hardest things of all the tasks that wine professionals are asked to do. When you are faced with flight after flight of wine, it takes experience, concentration, skill, and no small measure of physical stamina to do a good job. I'm particularly careful to give the first and last wines in the flight a proper chance (it can be easy to use the first few wines as a sort of aiming tool, as in darts), and I'm also careful to treat each wine on its own merits. For example, if you have had a run of poor wines, it's easy to overscore the next if it is halfway decent, whereas if you have had two excellent wines, you might be tempted to score the next a bit lower.

Style can be an issue too. I remember judging Chilean wines at a domestic competition about a decade ago with a group of tasters from the UK, most of whom had lots of letters after their names and classically oriented palates. It was tough going. Many of the reds were quite big, ripe wines, with lots of oak. So we tasted and tasted, and I got the impression that many of the wines that these tasters favored weren't actually particularly good, but were being rewarded for what they weren't—oaky blockbusters—rather than for what they were—simple commercial wines. This is a problem in judging wine blind: you can end up in a situation where some of the gold medal wines are nothing more than simple commercial wines. It's a case of the Emperor's New Clothes. (Thankfully, Chilean wines are a lot better now than they were ten years ago, although I do think the wines we judged weren't as bad as some of the tasters thought.)

I had a similar experience recently, judging wines in New Zealand. In New Zealand and Australia, wine shows follow a similar format. Each taster is seated at a table marked with glass positions 1–50 and three squares in each slot. Typically, the judges are faced with long flights of wine of a particular class, and they taste these flights over the course of an hour or so, pulling wines forward and pushing them back, and scoring them on a scale of 20 or 100. Most of the judges are winemakers themselves, and I've noticed that they have a tendency to reject anything a bit edgy, while the slightest hint of a "fault" compound is enough to get a wine booted. Praise, in the form of gold medals, is reserved for very competently made, clean commercial wines. I'm not doubting the ability of the judges; it's just that they are doing what they would in a winery when they are evaluating their baby wines: showing great vigilance. These, however, are finished wines, and they should be judged differently—from the perspective of the drinker. Many of the world's most sought-after wines would get a good kicking in an Aussie or Kiwi wine show, I suspect.

We must be careful not to praise wines for their lack of negative virtues.

32

Bright side story

*There's a lot of bad wine, but I'm not
going to worry about it. I'll just spend time
chasing the good ones. There are plenty
to keep me going.*

IT'S NOT so long ago that you could still easily
purchase very bad wine. Stuff you'd only buy because
it was all you could afford, and it contained a dose of
alcohol. The pain of drinking was offset by the gradual
numbing feeling of happy drunkenness. Over the last
thirty years, however, the average standard of wine has
increased substantially, and now if you go into a supermarket and buy one of the cheaper wines at random,
you'll almost certainly walk away with a bottle that you
can drink. It might bore you; it might taste the same as
lots of other wines; it might be confected and slightly
sweet—but I doubt it will make you wince because

of its awfulness. (In fact, some of the cheaper wines might be better because with the wafer-thin margins, oenological products and grape juice concentrate are too expensive to use in these wines.)

I've been critical in this book of bad wine, and by that I mostly mean boring, anonymous wine or tricked-up, spoofy wine. And there's a lot of it around. Of course, I'm glad we aren't still in the days of solidly bad wine, but the price of progress has been a degree of standardization, and now most wine is a bit bland. Which is sad.

But I'm not going to make moaning about bad things the theme of my writing, and I'm not going to waste time writing about such bottles. Nor am I going to let the bad, expensive, try-harder, point-chasing international wines distract me by their preposterous value proposition and terroir-denying affrontery. Instead, I will celebrate what is good in the world of wine.

Because, as wine drinkers, we are lucky to be living in times like these. We are not, of course, lucky that lots of very famous, excellent wines have leapt in price over the last decade, taking them out of reach of the likes of me and many of my friends. But we are lucky because never before have so many interesting wines been made. Go to pretty much any wine region and hunt around, and you will find keen, talented winegrowers prospecting for special vineyard sites and then trying to make wines that express these privileged patches of ground. This occurs in the new world, where pioneers have established great terroirs, but also in the old world, where a new generation of growers is emerging, shedding the complacency of their forebears, respecting the life within their soils, and making superb wines. For these brave souls, wine is a vocation rather than a job.

I get so excited when I'm introduced to or discover new producers who are making really good wines. Examples? Look at South Africa, with the Swartland revolutionaries, the Hemel-en-Aarde and Elgin explorers, and those revitalizing the talented vineyards of Stellenbosch. South Africa used to be a much less interesting wine country than it is now. And in the old world, the Loire and Beaujolais regions have both seen an explosion of really interesting producers, while Portugal continues to offer up surprises. And Spain is making some thrilling wines in areas that weren't previously associated with fine wine. There are two world-class producers on the island of Tenerife, for example.

I could go on. My point: there's never been a better time to be a wine drinker. So let's keep our focus on that.

33

No one drinks wine blind

How many times have you gone out for dinner and drunk wine without knowing what it was? Reduce wine to just a liquid in a glass, and it becomes much less interesting. But this is how so many of us taste professionally.

BLIND TASTING—SAMPLING A wine without knowing its identity—can be extremely valuable. It frees us from our prejudices. It gives every wine an equal chance to impress and seduce. When you are faced with a wine in a glass with no other information at all, it leads you to question it in the most honest way possible. And then there's the fun of guessing: where is this from?

When I'm drinking with friends, there is nothing more fun than to put a bag or a sock on a wine—or serve it already decanted—and then discuss it. We'll share our impressions, and then we'll embark on trying to identify exactly what it is.[1]

And then, crucially, we do the unveiling. At last, we see what we have been tasting. And we look at the wine again.

This is important, because once we know what the wine is, we can begin to understand it properly. The blind tasting is just a prelude. For interesting wines, the context matters. The details—the origin of the wine, its age, how it was made—all help us to interpret the sensory cues better. And of course, there is the context in which it is drunk. We can only ask the right questions of the wine if we know its context.

Stripping wine of its context reduces it to a liquid in a glass, at which point it is of limited interest. If wine is merely a liquid with certain sensory properties, we are all in big trouble.

Of course, wine is more than that. People drink wine for many reasons, and the taste of the liquid is but one of them, albeit a very important one. The

---

1. I heartily recommend Michel Tolmer's comic book *Mimi, FiFi, and Glouglou: A Short Treatise on Tasting* (Éditions de l'Épure, 2011). Originally in French, but now also translated into English, it follows the journey of three natural-wine-obsessed friends as they drink together. Tasting blind is their favorite thing. It pokes gentle fun at wine geeks, and Tolmer is an expert cartoonist.

drinking situation is vital to the enjoyment of wine, but so is our knowledge of what we are drinking. We may know a bit about the wine region; we may have some knowledge of the grape variety and the production process; and we may have an emotional connection with the wine for some reason, such as having visited the estate.

34

You don't drink the same wine at home

The situation is vital in the perception and appreciation of wine. This is why we pay a premium to drink good wine in restaurants. We are not drinking the same wine at home.

OFTEN IN THE wine world, people talk about the significant extra cost of drinking a wine in a restaurant. "It's an outrage," they say. "I'm paying three times retail—or more—to drink that wine. I could enjoy it at home for much less." For example, a restaurant has Jamet Côte Rôtie 2008 on the list. You have a case of Jamet 2008 at home. But the reason you pay extra for the wine in the restaurant is that you don't drink the same wine at home.

Think about airlines. People pay a significant premium to sit in a business class seat. I just searched for flights from London to San Francisco with British Airways for some random dates in May 2019, and the results I got were £1,400 for economy class and £7,000 for business class. That's a lot more money. What does this extra money buy you (after all, you are traveling on the same plane that gets you to the destination wherever you are sitting)? A more spacious seat that turns into a bed, and better food and drink. But you aren't paying more just to sit in that seat. If you were to take that same seat and put it in your home, it would be an entirely unremarkable seat. Your couch is probably more comfortable; and you've got a bed to stretch out in. Yes, in the plane you are paying in part for the extra room and greater comfort, but also—and perhaps most significantly—because it makes you feel more special. You have, by virtue of spending a lot more money, superior status to the people who turn right and file into the economy cabin. Airlines are very good at tapping into our status-seeking behavior, hard-wired into us through evolution, in order to make money from us.

No one who buys a business class ticket says, "It's an outrage, I can sit in a much more comfortable seat at home." It's clearly a silly thing to say. You pay the premium, or you don't.

There's a parallel here with drinking wine in restaurants. You don't drink the same wine at home.

As I've said, wine isn't just about the liquid in the glass. It is much more than that. When you go to a restaurant, although certainly you are paying for the food and drink, more tellingly what you are paying for is the ability to consume that food and drink at a certain time in a certain place. You are paying for an experience. To be sure, if you open a bottle of Jamet 2008 at home, you are having an experience, but it is different from the experience you would have drinking Jamet 2008 in a restaurant. Done well, a good meal is one of the most pleasurable things I can think of. The combination of the place, the company, the food, the drink, and the atmosphere all work together to create an experience. Yes, I enjoy drinking wine at home. But I pay a lot more, and gladly, to drink great wine in restaurants. And yes, some home-based meals beat any restaurant experience when you have good people in good humor, tasty food, and delicious wines. But my point still stands: you don't drink the same wine at home.

35

Beer is better than wine

*Many commercial wines are so deeply
manipulated that I'd rather drink beer.
Many in the wine trade are so disillusioned,
because they know that they are peddling
crap, that they lose their love for wine.*

WINE IS INTRINSICALLY interesting, and lots of people
are drawn to it. To work with wine is a dream for many.
But there's a side to the industry that is unattract-
ive—indeed, actually quite painful for people who care
about wine. As they peer behind the movie set, they see
that much of the talk about terroir, culture, beautiful
vineyards, careful winemaking, and the rural idyll of
wine country is in fact just a façade. As a result, many
of my colleagues in the wine trade have lost their love
for wine.

Those who take the wrong sorts of jobs find
themselves in a very different wine trade altogether.
They are selling cheap commercial wines, made from

large flatland vineyards that are kept clean of weeds by herbicide and sprayed with all manner of agrochemicals. The grapes, cropped at heroic yields, are machine picked, then transported to giant factory wineries where they are rushed through the winemaking process by vintners on speed dial with product company reps, made to a recipe with chemical help, then blended, sweetened, and sterile filtered, before being shipped off in a tanker and ending up on a supermarket shelf or on a restaurant list at a rock-bottom price.

Someone has to sell these cheap wines, but the stories that are made up about them, borrowing the language of artisanal fine wine, and trying to steal some of its glamor, are fairy tales. Many of those who sell such wines know they are acting in betrayal of what they know wine to be, and so they switch off that part of their soul out of sheer discomfort. They lose their love for the product they are working with, choosing not to talk about wine at all. It's all too painful.

Personally, I'd rather drink beer than suffer these dull, dishonest, tricked-about wines. Yes, some intervention is necessary in the winemaking process— though most is not. Good, traditional winemaking requires very little intervention and gives very good results. Is it ever necessary to add grape juice concentrate at the blending bench? A lot of winemakers think so. What about oak chips or powder? Copper fining? Bentonite? GrapEx? Powdered tannin? Commercial winemakers can give lots of reasons why they use these products, some of them quite good. But often it is only because they are in a rush, or they want to add make up, or they didn't get things right in the vineyard—perhaps because of the pursuit of heroic yields. Or maybe they just got scared by the stories the product reps told them and wanted to make sure nothing went wrong with their 40,000-liter tankfuls of wine.

There are some cheap commercial wines that I like, usually because of their frankness. But so many are artificial and dishonest, and if I'm faced with a wine list full of these dull, tricked-up products, then I'm not interested in drinking wine. Of course, lots of commercial beers are made in large factories with cheap ingredients. But it is often easier to find a good beer than it is a good wine, and usually a lot cheaper too.

If you are in the business of wine, and you are working with products you don't believe in, get out before it's too late and a bit of your soul dies, and with it any love for wine you have left.

36

Escape the small oak rut

Too many winegrowers are obsessed by small oak. Small oak—barrels and barriques—doesn't suit all that many wines. But it seems to be the default vessel of élevage. *It's a mistake.*

A COUPLE OF years ago I visited a *tonnellerie* in Burgundy—a factory where barrels are made from oak staves that have been seasoned for two or three years outdoors. It was fascinating to watch the process in action. First the oak staves are cut into the right shape, then the cooper takes metal rings and begins to form the barrel. The half-constructed barrel is heated over a flame to give the staves enough flexibility to be bent, and more hoops are added, with quite a bit of hammering, and finally we have the finished barrel shape. Some of the hoops are removed and the barrel is sanded, end discs are added, and a bung hole is cut in the middle of one side.[1] The barrel is tested for leaks, then finished, with some new hoops added in place of those used to hold it together during its construction. The finished barrel is a thing of beauty. And when you visit a winery, whether it is a small, rustic one or a big industrial one, the connection to the past is usually still present in the form of a room full of oak barrels containing fermenting or maturing wine.

The normal barrel size is 225 or 228 liters, commonly referred to as "small oak." This is a very convenient size because the empty barrels aren't too heavy and so are easy enough to move around, even in cellars that can't fit a forklift. This is the size you'll find in the cellars of Burgundy and Bordeaux, the two wine regions that many producers worldwide look to for inspiration. Typically, small oak barrels like these have a relatively short lifespan of five to seven uses. They are widely cherished when new because they add flavor to wine, though less is added with each subsequent use, until after year three or four the barrels are essentially neutral. But it's that first year of use that

---

1. I may have got some terminology wrong here. There is a whole lexicon of words associated with barrel manufacture, which is quite fun to explore. For the purposes of this chapter, I'm using simple terms.

adds the taste of oak that some people so love, as well as a bit of structure, and oak also helps fix the wine's color.

In the past, many regions didn't use small oak barrels at all. Instead the wines were made either in large concrete tanks or in large barrels of more than 1,000 liters, called *foudres* or *fuders* or *botti* or *Stück*, depending on the country. The larger volume-to-surface-area ratio meant that wine matured a bit more slowly in these vessels, and there was no flavor impact from the oak. In the 1990s, though, a trend for riper, richer, oak-flavored red wines caused many previously traditional producers in the classic regions to switch to small oak. Their large oak barrels were given the chainsaw treatment (they had often been assembled in the cellar, and there was no way to get them out without destroying them), and in came the small oak. In regions such as Barolo, where small oak had never been used before, a split emerged between traditionalists and modernists.

In the new world, small oak was adopted from the start, as many producers attempted to emulate their heroes in Burgundy and Bordeaux. The combination of sweet ripe fruit and oak flavor was appreciated by many.

In recent years, thinking has begun to change. While small oak is very convenient, some producers have questioned whether this one-size-fits-all approach is really the best option. There has been an exploration of alternative methods of élevage, including large oak, concrete eggs, and terracotta amphorae. This return to the past has resulted in some very exciting wines. In the classic areas that have seen a standoff between the traditionalists and the barrel-loving modernists, the situation has been more complicated, for many producers are in fact a bit of both. For some wines they'll use small oak, but for others they've gone back to the old ways and use much larger barrels. This variability is welcome: rather than sticking to a recipe, the best producers are experimenting and finding out what works best for them.

There's also the use of alternative woods—acacia, chestnut, and even cherry—which impart distinctive characters to wine aged in them. But these are less interesting than is a move away from small oak to a variety of containers, with the winegrower creatively—and skillfully—matching wine to vessel, much as a chef utilizes a range of pots and pans, all with specific uses. In some ways, we are moving back to the future with the use of terracotta and concrete. How the pendulum swings!

But let us not discount small oak altogether. In a recent tasting hosted by Languedoc producer Domaine Gayda, a group of trade and press were shown the same wine aged for eight months in nine different vessels, including oak of varying sizes, an amphora, a concrete egg, and even a polyethylene egg. I tasted all nine versions of this wine—a Syrah from the Roussillon—without knowing which vessel accounted for which wine. On this occasion, the wine I preferred was the one aged in a one-year-old French oak 228-liter barrel.

37

Segment or be damned

There isn't just one market, but many.
And different rules apply to the overlapping
segments that exist. We need to bear them
in mind or we will lose our way.

MANY DISCUSSIONS ABOUT wine, the wine market, and
"consumers" are formless. They go around and around
in loops. People make their points; others respond.
Yet there is a feeling that the conversation is circular
and repetitive. With nothing solid to grab hold of, no
progress is made.

This is because people think that when they discuss
the wine market, they are discussing something that
is coherent, singular, and whole. That simply isn't
true. The wine market is highly segmented, and
although there is some overlap, these segments are
largely separate, with different rules applying to the

different parts. All they share in common is a liquid called wine, but this liquid means different things to different people, is consumed in very different settings, and has different values attached to it.

Trying to define the segments, though difficult, is a worthwhile exercise, particularly if we wish to avoid these repetitive, formless discussions.

First, we have what is probably best referred to as commodity wine. It's the bottom end of the market, where all that people want is something red, white, pink, or fizzy that doesn't taste bad, contains alcohol, and isn't very expensive. This is most of the wine sold worldwide. It is your cheap milk chocolate or your instant coffee, to draw parallels with other product categories.

Second, we have wine as a luxury good, drunk largely by people who don't particularly care about wine—they wouldn't count it as a hobby or a special interest—but who have lots of money to spend and want the best.

This second segment overlaps significantly with a third segment: wine as an investment vehicle. And both overlap with the fourth segment: wine geeks. While all three segments fall under the banner of fine wine in its various guises, the wine geek has a particular focus: *interesting* wines (for example, from lesser-known regions or grape varieties), which may be inexpensive.

There's a big gap between the commodity segment and the fine wine/wine geek segment. The rules are totally different. One is not more important than the other, and indeed there may be some transfer between the commodity segment and the wine geek segment, with some wine geeks changing their purchasing behavior depending on the occasion, taking the occasional foray into the commodity segment.

When people refer to the "consumer" in discussions about wine, they typically fail to indicate exactly which "consumer" they are referring to. Usually it is the drinker of everyday, commodity wine, but that isn't made explicit. Most often, everyone is lumped together and mass confusion erupts.

One of the frequent sources of crossed wires in discussions about "consumers" occurs when the commodity wine drinker—the normal person—is meant. Wine producers are criticized for not appealing to the consumer, or for disregarding the consumer, by not making their product accessible enough, or by having difficult-to-interpret labels, and so on. But if a producer is operating in the fine wine segment, then criticizing them for not playing by the rules of a totally different segment is unfair.

Listening to some industry commentators, you'd think the wine industry is doomed. That it is failing, slipping into oblivion, and all because it has refused to modernize, it is stubborn, and it doesn't like consumers. Yet when I look at the segments of the wine trade that I find most interesting, and which I'm predominantly dealing with, I see a great success story. Across the world, many more interesting wines are being made now than in the past. New regions have emerged. There's also a tremendous diversity of wine being made, much of it compelling. Of the regions I've focused on—Portugal, South Africa, New Zealand, Australia, Canada—the wine scene is now *so* much more vibrant than it was twenty years ago.

My point? If we are to discuss wine, we need to segment the market. Wine isn't simply wine.

38

Balance is not always in the middle

Balance isn't beige. It isn't always to be found in the middle of two extremes. Rather, balance can be anywhere in between, just as the truth isn't necessarily found midway between two polar-opposite opinions.

BACK IN THE day, there was a craze for magnolia paint. I'm thinking the early 1990s, when I first became a homeowner. Whenever people renovated their homes, for some reason they chose to paint their walls magnolia, a sort of off-white color, and invariably they'd also choose off-white furniture and off-white carpets, all differing slightly in color but essentially creating the total beige experience. (The term *beige* is French, used to describe the color of unbleached wool: sort of off-white.)

Beige is also used as an adjective to describe people or situations that are essentially bland and boring in nature. Neither hot nor cold.

Well, balance isn't beige.

Take a journalist researching a controversial story with two sides to it. You might think the best way to approach it would be to interview people on both sides, and allow each to have their say. Fair, yes? Possibly, but it doesn't mean the resulting story is balanced. One side might have a minority, uninteresting, or possibly malign viewpoint that shouldn't be given as much weight. A journalist concerned with balance will attempt to convey this in their piece, rather than giving equal attention to both perspectives. Truth can sometimes lie at a pole.

As for wine? Sometimes a wine can be a little extreme, but still balanced. An intelligent interpretation of a site might require a wine with some edges—characteristics that stand out a bit. A balanced wine doesn't have to be right in the middle; it shouldn't be beige and inoffensive. For a wine to be true, it must sometimes be polar.

Consider a Mosel Kabinett Riesling. Is a great example of this wine one with moderate acidity, that pleases many and offends none? It could be that the best, most balanced example is one with high acidity, countered by some sweetness teetering on the brink of the extreme. Or take a great Côte Rôtie, with vivid peppery aromas and notes of meat and iodine: this, I'd argue, is a better, more balanced example of the appellation than a neighbor's wine with soft tannins, lush fruit, and no edges. Balance doesn't always lie in the middle.

39

We are on a journey; this is mine

Each of us is on our own journey. This metaphor applies to life as it does wine.

HOW DO PEOPLE discover wine? What proportion of non-involved consumers end up morphing into involved consumers? And how do people make this transition? These are all very interesting questions. The wine industry needs all sorts of consumers, but if we can understand how people first become wine drinkers, and then how and why some of them become more interested in wine, this could be very helpful.

All I really know is my own journey. Back when I was at university, I didn't drink wine much. I had come quite late to drinking any kind of alcohol, taking my first steps at about the age of seventeen, which by British standards is positively ancient. At that stage, though, I was already showing some geeky level of interest in beer, which I drank with a couple of slightly older friends after playing tennis or walking in the country. Initially, I found the flavors in beer a bit challenging, especially its bitter edge. But I overcame my initial aversion, and English cask-conditioned ales at the pub became my preferred drink. I remember being fascinated when I was introduced to Theakston's Old Peculier, a very strong, characterful ale (although by today's craft beer standards, it would not be considered particularly strong).

At university we drank beer, but occasionally we needed wine for dinner, and then I'd nip down to the local Gateway supermarket and browse the wine aisles. I was bottom feeding, trying to get the cheapest wine that would taste OK. Some of them tasted really bad: this was around 1989, and there were still lots of unpalatable wines around. Generally, I'd go for the one with the highest alcohol level in my price range.

I do remember the first wine that I tried that I found delicious, as opposed to just palatable. It was a Berri Estates Shiraz Cabernet 1989 (I think, or was it 1986?), and it would have been in the second year of my PhD, which puts it at 1990. The son-in-law of my landlady brought it round for Sunday lunch, and it tasted rich, a little bit sweet, and very smooth. Ah, I thought: there really is something to this wine thing!

After I completed my PhD in 1992 I moved to Wallington in Surrey, and for a while shared a flat with an older friend who'd been to both Oxford and Cambridge, where he'd been introduced to wine. He drank modestly (in terms of quantity) but well, and on Sunday evenings he'd often invite a few friends round and open one or two bottles. It was a revelation. I remember trying a Pouilly-Fuissé that actually had real personality, and this hooked me onto the flavor of good Chardonnay. I also remember being introduced to Claret: a 1982 Léoville-Barton did the trick. But the wine that impressed me the most was the 1991 Brokenwood Graveyard Shiraz, which became the first wine I purchased more than one bottle of (three, at £13.49 each). In my early stages of drinking wine, I grew to love some flavor motifs in certain varieties. I knew what I liked, but I didn't know why. It was a very raw form of wine appreciation, unencumbered by knowledge.

Two of my friends at the time, Michael and Jason, were also quite taken with wine, and we began buying and drinking bottles together. Michael had a copy of *Parker's Wine Buyers' Guide*—one of the early editions—and we began to look out for high-scoring wines in our price range. I loved the enthusiasm Parker showed in his write-ups: here was a true fellow wine geek. A happy place was finding his 86 A's—wines scoring 86, which at the time was a good score (before score inflation, maybe the equivalent of today's 90)—in the cheapest price bracket. We had new world palates, favoring wines with lots of flavor impact, but one quirky discovery that we enjoyed was Château Musar, which back in those days was inexpensive. We drank quite a few bottles of Musar. In general, though, we were disappointed with the wines that we bought at supermarkets. They had the same names as the wines we'd tried with our friend, but not the same flavors. I recall that another buying guide that came in handy was the *Sunday Times* guide authored by Robert Joseph, based on results from the International Wine Challenge, which was in its infancy then.

I had quite a bit of luck at the local specialist wine shop, called the Wine House. Operated by Morvin and Sue Rodker, the wines here were a bit more expensive than supermarket versions, but they tasted good. I was at the stage in my drinking career where I'd try something and find I really loved it, but I couldn't say exactly why, and I then struggled to replicate the experience. Evidently, I had a latent geeky interest, but it took people like my friend, and the nearby presence of a decent wine shop, to lure me in properly. They gave excellent recommendations. I remember them talking me up to a Chablis Grand Cru once, telling me: "Live a little!"

What else do I remember from the early days? Oddbins was in its prime in the early 1990s, and the eclectic selection on the shelves of this specialist chain was ideal fodder for a newbie geek. They sold the wines well, too, with enthusiastic staff and excellent wine lists. I also remember finding good bottles in the high-end specialist selection in the local Asda and Waitrose supermarkets. At the time, both stocked some interesting bottles. And then there were branches of Thresher and Wine Rack, with a few goodies in their offerings.

Fine wine was cheaper then, which made the odd foray into the classics affordable. I spent most of my time, though, exploring the new world, Australia being my first love. Indeed, my first visit to vineyards was in 1996, when I went to Melbourne for work. I hit the Barossa Valley, and we stayed cheaply in on-site caravans as we toured round Victoria. Our first-ever winery visit was at Charles Melton in March 1996. Australia was interesting back then, and the likes of St. Hallett, Rockfords, and Penfolds drew me in with their ripe but complex wines. Penfolds' Grange was still selling at £35 in the UK, before it was hiked massively along with the rest of the previously good-value Bin range.

I also remember my first encounters with *Brettanomyces,* although the time I didn't know what it was. A friend brought over a supermarket Crozes-Hermitage, and I found the farmyard aromas and strongly savory taste very offputting. I also encountered the same flavors in a Richeaume wine from Provence.

I began going to consumer tastings, which were quite memorable because at the time I didn't know to spit, so things started well and ended a bit blurry. I wrote down my impressions of the wines, taking the first step in developing a wine vocabulary.

By 1996 I was a proper wine geek, still very early on my journey but anxious to explore. And explore I did. With ready access to the World Wide Web, I started a hobby wine site on Geocities, which a couple of years later became wineanorak.com. I launched my current website in November 1999, so it's one of the oldest out there. I started blogging in September 2001, and mine is now the oldest extant wine blog. I learned a great deal from participating in the Wine Lovers' Discussion Group, a late-1990s US-based bulletin board for discussing wine online, and there was a golden period when professionals and amateurs alike shared their knowledge. It was at the right sort of scale for it to work, and very good-spirited. It didn't last too long, but for me as a keen newbie it was a superb place to make connections and learn.

In 2002 I started earning money writing about wine, as well as picking up advertising income from my growing website. In 2004 I got a book deal; in 2005 I got a national newspaper column; and in 2008, already in possession of a couple of Glenfiddich Awards, I quit my day job to work full time with wine.

So, is this typical of wine geeks? Are there lots of people out there like me, with a latent interest in wine that just needed triggering? Or do people simply get sucked into it, irrespective of any latent geekiness? I'd like to know whether there are useful strategies out there to get lots of people involved in wine. Or is it

just a few who potentially could go on to become serious wine drinkers? It may well be that some people
are more curious generally about flavor, and that they
will find their way in. But it could also be that some
people are put off because the only wines they get to
experience are uninteresting, and so they are never
more than uninvolved consumers who view wine
merely as a commodity. That would be a shame.

40

Don't expect the fish
to come to you

*In the wine trade we often behave
like unsuccessful fishermen waiting
for the fish to come to them.*

I'M NO EXPERT at fishing, but I can see its appeal. As
a child, I was fascinated by it, and it was a great thrill
when my father or grandfather took me out fishing. For a while my father had a small boat, which he
named *Bonne Chance,* and very occasionally he'd take
it out on the Thames, powered by a small, unreliable
outboard motor.[1] Once, when I was probably seven or

---

1. He was not the most successful boat owner. I remember once heading on holiday with the boat on a roof rack (I must have been very
young at the time; the memory is vague). At one point we had to turn
back because the boat was gradually sliding forward over the windshield, making driving perilous. And another strong memory is of the
fiberglass he used to repair the various holes that developed in this boat.
It had a very distinctive smell as it cured.

eight years old, I went out with him and my grandfather, and my grandfather fished off the back as we
puttered our way from lock to lock. As far as I recall,
he was not a great fisherman, because catching a fish
was a rare event, accompanied by high excitement and
a fair amount of celebration. This time, though, he
was lucky. And it wasn't just any fish, but a pike, a long,
streamlined, scary-looking fish with a terrifying array
of sharp teeth. We took it home, where the deceased
pike was kept in cold water in the bathtub until it was
time to eat it. (As an aside, pike makes for poor eating.
Too many bones, and a muddy river flavor.)

Aside from this triumph, I also remember many
enjoyable but fruitless fishing attempts. The key
problem may have been location: we fished where we
decided it would be most pleasant to fish, and hoped
that the fish would come to us. We should instead
have gone to where the fish were.

So often, winegrowers make wine in a certain
place and in a certain way because that's what they've
always done, and what their predecessors did. They
are fishing where they want to fish, and they are
just hoping something will bite. Some are lucky:
with a really good terroir in a well-known region, if
they don't mess it up customers will probably come

knocking on their door. But for many producers, this passive strategy doesn't work. Before making wine, it pays to do some research and find out where the fish are, so to speak.

Wine is a long-term game. Yet it is not uncommon for newcomers to winegrowing to make decisions almost on a whim. Those hasty decisions can come back to haunt you for a decade or longer. Think about the implications of planting a vineyard: you won't be getting a crop for three years, an income for four, and you won't really know just how good your site is for a decade. Plant in haste, repent at leisure.

Then you have to sell the wine. Making wine is hard, but in comparison to selling it, I am told, it's a walk in the park. I sense the pain of those who plant a vineyard as an early retirement project. They are likely starting a second career that is tougher than their first, and they'll probably be spending a lot more time than they were anticipating on the road trying to sell all those bottles that have tied up a lot of their capital. In a climate of oversupply, you really have to do something quite special to be noticed—and you really have to fish where the fish are.

41

Lead with your best

*Wine regions: try not to market yourself
through your cheapest, least impressive wine.
It doesn't tend to end well.*

BEAUJOLAIS AND CHIANTI are two wine regions where
the best wines have been held back by the regional
brand. Chianti became famous when the borders of
the vineyard area were expanded massively for commercial reasons, and the resulting flood of cheap, often
wicker-covered bottles was launched into the world.
The region became well known, but largely through
its cheaper wines. They filled a role: they were commodity wines that normal people could afford. But
they weren't very tasty. It's what the market wanted,
though, and Chianti is a memorable name. For the last
twenty years however, growers have been working hard
to tell the world that Chianti Classico, the original

vineyard area, is better and makes some wines we should be taking more seriously. But it may be too late.

Similarly, Beaujolais had this really smart idea: rush to the market with the new wines, dubbed Beaujolais Nouveau. In the days before southern hemisphere wines were readily available, this would have been the first chance to try wines from the new vintage. Released on the third Thursday of November each year, Beaujolais Nouveau became quite the thing, and the region became well known through this initiative. There was a problem, though. Hurried through fermentation and into bottle, these wines were cheap and unimpressive. Often, they reeked of bubblegum from the specific variety of cultured yeasts used to speed them through fermentation. The reputation of the region took a hammering, and it took a long time for Beaujolais to slowly claw its way back from this cheap wine rut. Only just now is it getting a second chance, with many artisan producers, farming properly and respecting their terroirs, making wines from the in-vogue Gamay grape that are causing the world to sit up and take notice. As with Chianti, though, the legacy of lead-with-your-worst remains, in terms of relatively low prices considering the quality on offer. It's the power of the regional brand.

How long does it take to recover from a bad reputation? The wine trade in the UK loves Riesling, but it is almost impossible to sell to normal people. That is because when the public at large was waking up to wine in the 1970s and '80s, sweet, cheap, poor-quality Liebfraumilch defined German wine. Most of this was made from the Müller Thurgau grape not the more highly regarded Riesling, yet the mental link between Germany, with its tall flute bottles, and bad cheap sweet wine was firmly etched in the public mind. To this day, Riesling is guilty by association, and selling wine in those flute bottles—however good it is—is very hard work. Alsace, which uses a similar bottle shape, is suffering from this bottle contagion.

So if there is a lesson from this, it's this: you get one chance to present yourself to the world. First impressions matter, and you need to lead with your best. Sometimes, if you are lucky and persistent, you may get a second chance and a bad reputation can be overturned. But it's rare.

42

Beware the wine consumer champions

There is a new breed of consumer champion loose in the world of wine. Their message may be seductive, but their anti-elitist spiel cloaks a dangerous message.

A BAND OF vocal marketers, commentators, and "thought leaders" are outraged by the wine industry's loathing, contempt, and disregard for "the consumer," which leaves large swaths of the drinking population feeling criticized, belittled, and vilified.

Angry with the wine trade, they have aligned themselves with "the consumer," a largely undefined population of hurting, neglected folk who represent everyone who drinks wine apart from the wine trade.

In this new narrative of wine, the cast and plot are simple.

On the one side we have the baddies. This shady crowd consists of the wine trade at large, and anyone with wine expertise, or who finds wine interesting, and enjoys the culture of wine, fine wine, natural wine, sharing interesting bottles with geeky friends, small-production wines, wine books, and wine education.

On the other side we have goodies: the consumers. These are people who don't know much about wine, don't want to spend much on it, but really enjoy their wines and get a lot of pleasure out of them, who drink with friends, who are happy most of the time. They are simple, joyful folk.

It's clear which side any right-thinking person would be on, right?

As for the plot, it goes something like this.

The wine trade hates consumers. It tries to make wine complicated and difficult out of sheer spite, to keep consumers away. And when these baddies see consumers having fun, they try to ruin their plea-sure by criticizing their choices. The great tragedy is that instead of advising consumers simply to find what they like and endorsing their choices, the wine trade often suggests that the wines the consumers are drinking are of poor quality, and that by spending a bit more and learning about wine, consumers could be having a better experience. Such an attitude is shocking, sickening even, say the consumer champions. It leaves the consumers feeling demeaned, belittled, and, of course, vilified. Always vilified.

But it's OK, for into town ride the heroes of the hour, the consumer champions, in a frenzy of faux outrage. They are here to help. Thank goodness! Their rescue strategy is two-pronged. First, they reach out to the consumer, placing an arm around their shoulder and speaking soothingly: "We're not like the rest of the wine trade," they say. "We get you. We're on your side. All this business about wine complexity? It's nonsense. There's nothing to see. Just enjoy the wines you're already drinking. They are great!"

Next, they berate the wine trade, especially anyone possessing expertise. Fight back! Their special talent: exposing structural problems in the wine industry. Production is too distributed. There's a mismatch between the scale of production and modern retail, which means the route to market is tough for many producers. Many small family businesses struggle to make money. It's expensive to make really interesting wine, and this puts it out of the reach of many people's budgets. And wine is unbelievably complex, yet many producers make little effort to help reduce this complexity. Finally, they argue, most people just want a glass of wine that doesn't taste bad and isn't too expensive. All these points are true. But the consumer champions offer no real solutions to these problems. They just beat the wine trade up about them.

Consumer champions love innovation. The wine trade, they say, resists innovation, and that is one of the reasons it is in trouble. Why does it resist? Because the wine trade hates consumers, and consumers want innovation, so the wine trade stubbornly resists just to spite them, which is shameful and saddening and brings us close to tears. So the consumer champions celebrate any innovation, however crazy, inappropriate, or laughable.

Remember: the consumer champions are all futurists. They anticipate a future where you can have a device on your phone that reads your DNA, detects your emotional state, chooses a wine that matches both your biology and state of mind, creates it in three minutes by chemical synthesis, and then delivers this personalized wine by drone in a further two minutes in novel recyclable packaging that sequesters carbon dioxide from the environment as it self-destructs once you've finished. Don't criticize this idea: in 1979, when you were given your first digital watch, you'd have found the idea of an iPhone ludicrous. This is progress, and all progress is good, to be welcomed and be claimed as our own because we saw it coming.

But it's when it comes to attracting new customers that the consumer champions get most fired up. They want to help the wine industry to grow, sell more wine, and above all recruit new customers. They will come and speak at your conference for a modest fee, tell you off, and then tell you how to win new customers.

The solution? Strip wine of its complexity. Get rid of all the experts with their annoying expertise. Make wine taste nice again. Sweetness is helpful here, because young people have simple tastes and want things to be sweet and easy. Young people are scared of wine, and would rather drink sodas, alcopops, mixed spirits, fruit ciders, and beer.

So we need to encourage the wine industry to make simple, sweet wines that taste more like soft drinks or fruit coolers than wine. Add flavorings to wine? Why not? Anything that makes it simple and easy for young people, with all their limitations, to enjoy. And if the young enjoy these concoctions, don't tell them they are wrong, because they like what they like and they are the ultimate arbiters of taste. Who gets to decide what is a good wine and what is a bad wine? It's you, of course, the consumer! All we want is to help the wine industry sell more wine to more people.

My response? I think the consumer champions are well intentioned. But they are misguided. They identify many of the problems in the wine industry, but offer few solutions, and some that they do offer are actually dangerous for the wine trade.

There is a structural issue in the wine industry, with a gap between the scale of production and the scale of modern retail. Wine is an incredibly fragmented business, and few people make much money out of it. Consumption in traditional wine-producing countries is going down.

But wine is necessarily complex. It is different from other drinks. And if you segment the industry, you see tremendous success stories alongside the tales of woe. There has never been such global interest in interesting wine, nor has so much interesting wine been made. When I travel the wine world, I see lots of engagement with younger drinkers. I see regions transformed as young *vignerons* take over and make interesting wines from well-farmed vineyards.

Either the consumer champions don't see this, or they choose to ignore it. They champion wines that have few of the qualities that make wine interesting and unique. They celebrate processed wine. They focus on people making poor wine and struggling to make a living, tell the wine industry how badly it is doing, and take an anti-expert stance.

Look at bread, or coffee, or chocolate. There's instant coffee, there's white sliced bread, there's mass-market milk chocolate. As a teenager and young adult, I consumed all of these. Now, as an adult, I realize that none are good quality. They serve a purpose, to be sure, and many people enjoy them. But it's insane to expect commentators in those categories to extol their merits. You don't expect restaurant critics to write about McDonalds or Pizza Hut. It's the same with wine.

I actually taste a lot of commercial wine—more than most wine journalists. It's not just a diet of the fancy stuff for me. And most mass-market, processed wines are objectively poor. That's simply the way it is. If you ask me, I will tell you. Sure, millions of people consume them happily, but some of these wines are evil, more likely to put people off wine than recruit them to the category. They sell because they have distribution, not necessarily because people enjoy them. Rather, it's all they can find to drink within their price range and at the point of purchase.

This is not something we should celebrate. It's perfectly appropriate to tell people that they can drink better wines, ones that are culturally rich, made from vineyards that are sustainably farmed, and that could actually enhance their lives. I wouldn't walk into their living room and tell them that, but they aren't going to read what I write unless they are looking for it.

Because this whole consumer champion line—the idea that people are being shamed or upset or put down (or, of course, vilified) by experts saying that they could be drinking a better product—is just a silly myth. People drinking sweetened-up reds and enjoying them will keep on drinking them. They don't read about wine—it's far too abstract. There are precisely zero people out there who are feeling offended right now because wine experts are saying that the wines they drink aren't very good.

There may, of course, be some who feel *insecure* about their choice. They suspect they may be drinking crap wine, but that's because they sort of know that better wines do exist. It's like me meeting a coffee expert while holding a Starbucks flat white in my hand. I know there are better options out there, and so I might feel a bit embarrassed, but often I'm perfectly happy with my lesser options—all of which are better than the instant coffee I used to have as a student. There's no need to blame the coffee industry for having quality tiers. And I don't feel vilified.

The consumer champions mean well. They often sound plausible. And they will try to shut you down if you disagree with them. So don't be taken in by their simplistic narrative, which ignores complex reality. It is in the details—and in segmenting the market— that the truth is found. Wine is far too rich, complex, and diverse to be understood through such a simple worldview, or to be rescued by the false salvation they offer.

43

Winemaking is not chemistry. It is biology

We often think of wine in terms of its chemical composition: acid, alcohol, aromas, and so on. We are better served if, instead, we think about it as biology. It's the microbes that transform grape juice into wine, and they should be more central in our thinking.

IT'S NORMAL IN wine and wine science to think about wine as a chemical solution, consisting of a number of different chemical entities, some of which have taste and smell properties. This is a useful starting point, but it leads us to believe that winemaking is chemistry, a series of interventions designed to end up with a desirable mix of chemical components.

For example, we want to have a backbone of acidity and/or tannin. We want the wine to have the right pH, so it is stable. We want some fruit characters, so we act to protect the must and later the wine from too much oxygen. We want some harmony in the wine, so we allow lees contact, or we age in barrel, or we do both. We might add a range of enological products, to protect and enhance the wine.

But in truth, wine is about biology. It begins with the grape vine, a climbing plant normally at home in the woods, but which we have domesticated. Considering the vine in its original habitat gives us great insight into viticulture. We see it as a plant that is used to establishing itself in an environment where there is already a lot of competition. It has to interrogate the soil for water and nutrients in a space that is already full of other roots. This makes the vine an expert scavenger, and helps explain why it flourishes in otherwise uninviting, bony hillside soils where other crops offer little reward.

The vine also uses other plants as support, climbing up by means of tendrils. With the exception of bush vines, which succeed only in some climates and with some varieties, vines therefore need some form of trellising or other support. In some rare cases, vines used to be grown up trees as support, which made pruning and harvesting somewhat tricky, best suited to those good with ladders.

In its native environment, where the vine reaches a hole in the canopy, the sunlight exposure on the bud causes a change in morphology: instead of producing tendrils, next year's bud—present already in embryonic form—decides to make a flower cluster. Then when grapes form, the birds can see them: remember, grapes are for the birds, and the color change and accumulation of sugar signal that the seed is ready to be dispersed. (In nature, all grapes are dark-colored; whites emerged only after domestication.)

Grapes are picked, and the remarkable thing about them is that they have all that they need in them and on them to make wine. Their skin is populated by wild yeasts. As detailed in chapter twenty-eight, these yeast species work in shifts to carry out fermentation, with different species acting at different stages. The bulk of the work is carried out by a single species, *Saccharomyces cerevisiae,* and recent research has shown that while this is present in the vineyard at only very low levels—so low that for a long time wine scientists doubted that it was present on the grapes at all—there is enough there that after a few days of fermentation it dominates.

The curious thing about *S. cerevisiae* is that it has developed a special evolutionary strategy to carve out its own niche. After all, there are lots of other microbes that would covet the sugar-rich milieu of ripe grapes. So *S. cerevisiae* goes in for some habitat engineering. Instead of using up the sugar in the most efficient way, by respiring it, it chooses a less efficient method, which we call alcoholic fermentation. This method releases less energy than respiration, but the yeast willingly takes the hit because the by-products of fermentation—alcohol and heat—are too much for its competitors to cope with, so it gets all the sugar to itself. Sometimes the toxic environment thus created is too harsh even for *S. cerevisiae*, but of course, it was never intending to make wine!

In addition to alcoholic fermentation, we have other microbes in the mix, including a number of spoilage organisms (bacteria that can produce volatile acidity and mousiness, and the rogue yeast *Brettanomyces*), as well of course as the bacteria involved in malolactic fermentation.

It's important to think about wine in terms of biology, and not chemistry, for this changes the way we then approach it. The philosophical underpinnings of wine are important because they shape our actions. If we think of wine as chemistry, then the winemaker becomes God and all agency has passed to her or him. Instead, by considering it to be about biology, winegrowers shift their approach and see themselves as overseers of a natural process where the skill and knowledge they have accrued permits them to intervene only when it is essential. As wine drinkers, we recognize the specialness of wine as a gift from nature and a living thing, something to be savored with thankfulness and reverence.

44

Stop trying so hard
and just be yourself

*Sometimes I wish cheap wine didn't try so
hard. Why can't wine just be wine? Do we
need the sweetness, the oak products, the
polishing, the fake aromas?*

WHEN I WAS a student, entering the second year of my
PhD, I bought a car. It cost me £1,300, which was a
princely sum at the time. Nothing at all fancy: an old
Vauxhall Astra Estate in sky blue (though I'm sure the
color had a much fancier name: car colors always do). I
bought it from a local repair guy with a good reputation—he was honest, I was told, which for someone
selling a car isn't always a given—and for part of the
deal he repaired a rusty rear wheel arch. (In those days,
cars rusted a lot more than they do now.) Initially, the
car was a little unreliable, with frequent uncertainty as
to whether it would start. But I changed the battery,
and later the alternator, and later the windshield wiper

motor, and overall it was a good car, loyal and humble, and I grew quite affectionate toward it. I particularly remember the drive through Epsom Downs on my way home from seeing my then-fiancée in Wallington, Surrey (I was studying in Egham). It's strange how some routes stick with you.

Sometimes you need a car, and money means it can't be a fancy car, and that's OK. But I have never understood people who buy cheap cars and then pimp them up to look all fancy. It's like lipstick on a pig. Unnecessary, maybe a bit funny, but mainly just silly. Putting fancy alloy wheels on a beat-up old car is just daft. Go-faster stripes are always a bad idea. And as a general rule in life, don't spend three times the value of your car on a snazzy sound system, with the entire boot (or trunk) filled with a subwoofer. Oh, and lights are for the front and back of the car, not for a strip around its perimeter at floor level.

The connection with wine is obvious. Winegrowers want to make the best wines they can. But nature isn't egalitarian. Not all vineyards are equal. Some are capable of making great wines, while others can only make humble wines. Others shouldn't have been planted at all. And beyond this, vineyards have different talents, regardless of their quality potential.

It seems crazy to try to grow everything in the same vineyard; isn't it better to identify a vineyard's talent, and stick to that?

Once a winegrower gets the grapes into the winery, things can go very badly wrong. There was a period when we had the cult of the winemaker: talented winemakers, it seemed, could turn any grapes into something special. There are now many tools, both physical and chemical, available to winemakers that feed into this narrative. Yet wines made this way aren't interesting. Far more desirable is an honest, humble wine, rather than a wine that isn't being itself.

45

Don't be an all-rounder;
be a specialist

*No one is good at everything, and we
waste a lot of time and energy focusing on
what we aren't especially good at, when
instead we should embrace our talents,
once we have found them.*

"WE DO LOTS of things quite well" is one of the worst
marketing messages ever. But it's one that many wine
regions and wineries lead with when they come to tell
their story.

People are all different. Yet as we grow up, society
puts a squeeze on us, encouraging us to conform.
The books we read and the movies and TV shows
we watch present us with ideals and standards, ways
of behaving and fitting in. Our culture has a message
that we internalize unconsciously, and it affects

how we feel about ourselves and our place in the world.

At the same time, though we are aware that we're all different. Some kids are good at some things, others at other things. And society rewards those who do well, with stickers, medals, certificates, and prestige. This carries through to adulthood, and to some extent it seems that life is one big popularity contest where success is fetishized.

Discovering ourselves—finding out who we really are—and then working out what our specific talents are is a difficult but important journey. We don't always have the good fortune of doing something we are good at, or enjoy, for a living. Sometimes our success can propel us into careers that reward us in terms of prestige and money, but that leave us unfulfilled and miserable. There are many people who feel unsatisfied because they should, according to society, be happy, but they have never really found themselves or what they were meant to do.

And those of us who do find ourselves often have difficulty focusing on what we are good at and accepting that there are some things we simply won't ever do very well. We waste time and emotional energy trying to manage our flaws and weaknesses, seemingly unaware that very few people are all-rounders.[1] How liberating to realize that it's fine that we aren't good at some things, that it's not a source of shame or embarrassment. Instead, we can focus on our strengths!

Wine regions have a similar journey of discovery. The world's great regions are not all-rounders. Wisely, they have focused on their talents. Look at Champagne. It's a region that almost exclusively makes bottle-fermented sparkling wine, at which it has been tremendously successful. Burgundy's fame rests on two varieties: Chardonnay and Pinot Noir (although there is the tiny but wonderful niche of Aligoté). The Mosel is all about Riesling. New Zealand's Marlborough region has hung its hat on distinctive Sauvignon Blanc, although it does grow some other varieties. And Provence is famous for the color of its wine: the (increasingly pale) pink of rosé.

The temptation for new world wine regions is to be all-rounders. Wineries often start by supplying local

---

1. An exception to this advice would be to those whose flaws are obvious, unpleasant, debilitating, and upsetting to others. For example, if you have an anger problem and shout aggressively at your partner, children, and neighbors, then get some therapy. Or if you are habitually late and keep missing things, work on this a little. I'm not handing out free passes for lazy, mean, or nasty people here.

markets, whose customers want a full range of wines. And the winegrowers typically want to experiment, so plant a range of grapes, in part out of curiosity, in part because they are hedging their bets. As a result, many emerging regions find it difficult to arrive at a simple, coherent marketing message. Consider, for instance, Hawke's Bay in New Zealand—a region that makes some excellent wines, but that is often overlooked because it has struggled to define clearly what it does. To my mind, Hawke's Bay should tell the world about its excellent Chardonnay and Syrah, two varieties that, grown in the right places, are capable of greatness. But some winegrowers in the region are proud of their red Bordeaux blends, so this gets added into the marketing spiel. Others work with aromatic white varieties. A few even grow some decent Pinot Noir and Sauvignon Blanc. Quickly, the regional message becomes overcomplicated and diluted in its impact.

Compare this approach with two other regions in New Zealand: Central Otago, which has sprung to fame as home to some of the new world's best Pinot Noir, and Marlborough, home to a distinctive style of Sauvignon Blanc that has taken the world by storm. Travel to each of these regions and you'll find other varieties, many of which are making excellent wines. But as regions they have been wise to lead with simple—and simplified—marketing messages that have had real power.

The thing is, some consumers will enjoy the oddities or minority wines made in regions with specific specialisms. But others only have the capacity for a simple, concise, and clear marketing message. As a region gains fame through a tight, focused marketing message, those who make wines that aren't covered by this message will generally benefit also. Everyone wins, even if initially it may seem that some are being left out.

One example that I have been quite close to has been the emergence of Great Britain as a source of world-class sparkling wine.[2] Until fairly recently, English wine was a bit of a joke, dominated by small producers in a marginal climate making slightly

---

2. Until recently we all referred to "English sparkling wine," but there are some vineyards in Wales, too, and they were feeling left out. So we now refer to the wines of Great Britain. There's a reluctance, however, to use the term *British* in conjunction with wine because there still exists a category called British wine that is a horrible historical artifact made in the UK from grape juice concentrate.

mean, tart wines that tasted of nettles, elderflower, lemon juice, and rainy, disappointing summers. Now, though, people have realized that it's possible to make very good sparkling wines here.[3] Indeed, some big players are involved, and the industry has professionalized, and there's a lot of excitement— although a few naysayers look at the rapid expansion of vineyard area that has taken place and predict a glut followed by a price crash, and maybe the end of civilization, or something of the sort.

This is a great marketing opportunity. Here we have a chance to tell the world about what we do best. It's a simple, clear message, and thus carries great power. The wines are really good and back the message up brilliantly. Quality is reassuringly consistent. Yet there are still those who want to talk about English still wines, most notably Bacchus. Now, there are some good still wines, and I'm not suggesting that winegrowers shouldn't persevere with still wines, nor that there shouldn't be continued experimentation. So far, though, they are not internationally competitive, they are inconsistent, and I don't really want to

3. For the curious, I can recommend Nyetimber, Gusbourne, Hambledon, Ridgeview, Camel Valley, Hattingley Valley, Exton Park, Rathfinney, and Coates & Seely as good producers to start with.

drink many of them. It would be a tragedy if a beautifully clear, powerful marketing message were diluted by adding, "and we also make some interesting still wines, such as Bacchus." Once you have found your talents, focus on them.

46

Beware the time lag

*There's some truth in the expression
"You reap what you sow." But it doesn't
happen immediately. There is often a time
lag, and this frequently catches people out.*

A MISTAKE PEOPLE commonly make is to assume that
how well their business is doing reflects their current
performance. I know I do this with my own work. As a
wine writer—or more accurately, a wine communica-
tor—the reputation I have and the work that is coming in are not the result of my current performance.
There is a time lag. I'm reaping what I sowed some
time ago. In fact, the sowing analogy is a good one. No
one thinks it is surprising that there is a delay between
sowing seeds and harvesting a crop, but this time lag
effect is frequently ignored when it comes to looking at
performance metrics of businesses.

There is often a significant delay between action and results. Consider a winery. Of course, sales do reflect the current performance of the sales team, and it is by meeting sales targets that they tend to get rewarded. But the reputation of a winery is built up over many years, and there is always a delay between the work in the vineyard and the consumption of the wine by customers. Sometimes the delay is just one season, but often it will be more than a couple of years. So in addition to the work of the sales force, the winery's activities one or more years ago influence current sales.

If we are to look at performance metrics, we therefore need to have some idea of the time lag between performance and results. If a winery starts underperforming, by the time it becomes apparent in graphs another couple of years may be required to reverse this slump. The delay between sowing and reaping can be quite long, and if we are always reacting only to metrics, we might fail to understand what is really going on. We need to do more than just look at the graphs if we are to avoid underperformance.

47

First impressions matter

First impressions matter much more than we realize. The first impression a wine makes on you is vital, and this is something winegrowers can use to their advantage.

FIRST IMPRESSIONS COUNT. I'm increasingly convinced of this. The first line of a novel: it matters more than it should. The opening scene in a film. The intro to a song. A stand-up comedian's first joke. And the first time you meet someone: this is especially important.

The first impression you give when you meet someone—those first few seconds—creates the filter through which all subsequent interactions are processed. I can think of some people I get on really well with, and in part I put this down to a very positive first interaction. Subsequently, each meeting carried with it that positive vibe. We view following interactions through the lens of that first impression.[1]

We rely much more on our instincts than we realize, and I suspect that the very first few seconds after meeting someone can shape future encounters. To a degree, at any rate.

Of course, bad first impressions aren't deal-breakers; it is possible to recover from one, though it can take a long time. Part of the reason for this is confirmation bias, which affects the way we deal with information. If someone made a good first impression on us, then we have the internal story running in the background that this person is good. Our confirmation allows us to ignore evidence that doesn't fit with our narrative (i.e., evidence that the person is bad), and to marshal evidence in support of the internal story we run (that the person is good).

For wine marketing, first impressions are also critical. You really only have one chance to tell the world about what you do. You need to take that chance and make the most of it. So it's worth waiting for the right moment to tell everyone your story, and you need to tell it well.

---

1. Of course, there could be a confounder here: it may be that I made a good first impression with them because I was going to get on really well with them, and this is something we realized when we first met. Allied to this: do people have a sort of "spirit" that you recognize immediately? Sometimes I think so.

When I taste wine, the first impression is also important. Packaging matters. Expectations matter. The prejudice that comes from sighted tasting is important because it influences perceptions, and you can use it to your advantage if you are presenting wine to people. If people expect a great wine-tasting experience before they taste a sip, it is more likely to be one.

48

Play the long game

*Planting a vineyard is an act of faith:
it will be likely three years before you get
your first crop, four before the wine is ready,
and you'll only really know how good the
wine is after ten years or so. It pays to play
the long game with wine.*

I ADMIRE PEOPLE who plant trees. Planting a tree is
a great example of someone playing the long game,
thinking far into the future. On one of my favorite
walks, I've come to know a few venerable oak trees that
are perhaps 150 years old. They are grand trees, full
of wisdom and beauty, and I sometimes think of the
people who planted them, now long dead. And here I
am, one or two generations later, enjoying them. Oaks
are important to wine, too; the barrels that fill the cellars of wineries worldwide came from forests planted
by previous generations. I love this continuity.

In our lives we frequently must choose between
immediate pleasure and deferred gratification.

Planning ahead and saving for the future is a basic life skill, and those who are unable to do so usually suffer in today's society. As a nineteen-year-old, we might decide to continue our education and delay our entry into the world of work, and we cast envious glances at our friends who chose the second route and have their first paychecks. But we know that our choice will likely be rewarded some distance down the track.

Similarly, solid, steady performance often reaps a greater reward than rushing ahead. The hare sets off on the race faster than the tortoise but becomes complacent and begins to take it easy. The tortoise's steady progress ends up winning the race.

Wine isn't an industry where short-term plans are going to work. You only start a winery if you have a long-term business plan and are adequately capitalized, so you can do things right. When people are short of money, don't have a clear vision, they make poor decisions that will limit the future growth and strength of the business. Every decision needs to be checked against the visions and values of the project, in order to keep the project growing healthily. This is playing the long game.

49

For most wines, glass bottles make no sense

In many classic wine-producing countries it used to be normal to visit a winery and buy wine directly from the tank, filling up your own container. This is a good thing: many wines don't need to be bottled.

HOW DID YOU spend your childhood holidays? I have three siblings, and my parents took the four of us camping every year. This was quite brave: on our first camping trip, in France, my twin sister, Anne, and I were just over four; my younger brother, Arthur, was a newborn; and my sister Hester was somewhere in between. We went to somewhere on the Atlantic coast, and it rained solidly for two weeks, with only the occasional glimpse of the sun. Eight-millimeter home movies show flickering images of us all bundled in coats, avoiding rain showers, and playing with the snails out enjoying the weather.

My parents learned their lesson, and in subsequent years we headed to the Mediterranean coast. Provence was next, but there there's the risk of the Mistral: getting sandblasted on a beach isn't so much fun. And even then, back in the late 1970s, it was very busy during August. After a few years they settled on Spain; first Tarragona, and then a campsite near Valencia that was to become a perennial favorite. When my father went freelance in the early 1980s, the holidays became longer. One mammoth trip lasted six weeks, our entire school summer break. It was hot and sunny, and because it was even hotter inside the tent, we were outside all day. My parents loved to relax on the beach and drink wine, and then in the evenings, they loved to cook and drink wine. And they frequently visited bodegas in town where they'd take their own containers and fill them with inexpensive wine straight from the tank.

In many wine regions, buying *en vrac* like this (to use the French term) was the norm. It made sense: why bottle a wine when it is to be consumed in the very near future? It's a waste of glass, plus there's something wonderfully visceral and immediate about filling your own vessel with an honest, inexpensive wine straight from tank. The wine doesn't have to taste all that great: that isn't the point. It is a daily staple, usually relatively light in alcohol, thirst-quenching and digestible, true and affordable. Wine for the people.

There are many wines that really should never see a bottle. Of course, if you live some distance from a wine region, the wine must be transported and kept in good condition during the journey. But there are alternatives to bottles, and I'm all for them, whether it is bag-in-box or bladder pack, or the new thing in restaurants, wine on tap. Two wine-on-tap systems are widely used in the trade nowadays: Petainer and KeyKeg. Both are fabulous. The restaurant can either pour by the glass from the tap, or top up carafes or bottles (which are then washed and reused). It saves a lot of glass, and the wine is always in good condition. I welcome moves away from selling inexpensive wine in bottles. It really isn't necessary.

50

Fewer follies

*One of the things making life difficult
for wineries is that some of their neighbors
are follies: they don't need to make
money because they aren't proper,
stand-alone businesses.*

THE UK IS littered with architectural follies: grand-looking, generally nonfunctional buildings, often placed strategically in the grounds of a fancy stately home. Popular in the eighteenth and early nineteenth centuries, follies were a display of wealth, an affectation to satisfy those with too much money. Typically, they were designed to look like old castles or even classical temples. Virginia Water in Windsor Great Park (which I knew well from my student days) even has a mock ruined Roman temple, complete with fallen masonry and columns. As a kid, I used to love these faux vintage constructions. Anything with a few turrets that half resembled a castle would get me excited. I well remember sketching Mow Cop castle, a hilltop folly on the boundary of Cheshire and Staffordshire that we could see from my aunt and uncle's home.

The wine world also has its follies. These are wineries that, in all likelihood, will never turn a profit. They exist as an extension of a wealthy person's ego: "I am rich beyond the dreams of avarice, so what else can I do with my money? I know: I'll plant vineyards and build a winery." Wine is, of course, a capital-intensive business, and some people have enough money that they see bearing the losses from a nonviable winery as the entrance ticket to the good life, as a vigneron of sorts.

The problem is, these follies muddy the water for the rest. The influx of wealthy wannabe winegrowers pushes up vineyard prices. It's hard to compete with someone who has effectively bottomless pockets. Expensive vineyard land can be a problem for those who need to make a living out of wine; it makes acquiring new vineyards impossible, and pushes grape prices into the stratosphere. The vignerons find themselves sitting on land that is worth a fortune, which has implications for inheritance tax, making it very difficult for the next generation hoping to strike out on their own.

In 2019 I interviewed Jay Boberg, who is making wine in Oregon with Jean-Nicholas Méo under the label Nicholas Jay. He explained how after a successful career in music—he cofounded IRS Records in the late 1970s—he developed a love of wine and wanted to make Pinot Noir. "I wanted to do something that wasn't going to be a folly," he told me. "If you are starting a winery today in California, you are probably someone who has a lot of money and you don't care if you ever make any money." He adds, "Maybe you are someone who likes building things, but at that stage in your life, where you are economically means that whether it is profitable or not is not your main criterion. I don't want to be that guy." Now, five vintages into his Oregon project, Boberg thinks they will be cashflow positive this year, or at least next. "I'm extremely proud of that."

While one the one hand it's good that rich folk are attracted to wine and enjoy it, and plow cash into the industry. But these follies do mess up the economics of some of the more popular regions, and that's regrettable.

51

Celebrity wines, no thanks

Please, can we put a stop to wines "made"—or endorsed, which in many cases is the same thing—by celebrities?

YOU MAY HAVE expertise in one field, but it isn't always transferrable. Celebrities have enjoyed success, and wine attracts people with money. Sadly, this often results in celebrity wines: famous people want a wine label with their name on it. The result is almost always embarrassing.

I remember as a kid being obsessed with cricket. It must have been at about age fourteen when out of nowhere emerged a deep love of one of the most bizarre of all sports.

For those of you who don't live in one of the Commonwealth countries (cricket is enthusiastically played mainly in countries that at some stage have had a close connection with England, including India, Pakistan, Australia, the West Indies, South Africa, and New Zealand), some explanation is needed. Cricket is a team sport, but, a bit like baseball, involves individual contests and performances. A bowler bowls (not throws: the delivery action must involve a straight arm) a hard, leather-encased cork ball toward a batter (now the preferred term to *batsman*). Importantly, the grass-covered pitch comes into play because the bowler aims to pitch the ball in the right place in front of the batter to make it difficult to hit it, so in a sense the game is all about the grass, its preparation and quality, which can drastically affect the outcome of the game.

Suitably encased in protective gear and with a four-inch-wide piece of wood in his (or her) hands, the batter has to, first, keep the ball from hitting three wooden sticks known as stumps, but also try to hit the ball far enough that he can run to where his partner is standing in front of a second set of stumps some twenty-two yards distant, in order to score a run. Hit the ball farther, and more runs can be scored. But if the batter hits the ball in the air and it is caught, he is out. To explain the rules of cricket in any more detail would take a whole book. Unless you have grown up with cricket, it's unlikely you are going to acquire a taste for it in later life. I should add, that, quite wonderfully, the top form of the game involves "test matches" that take five days to play. It is a game rich in tradition, and I find it quite compelling.

So as a child I became obsessed with this game and would frequently travel to Lord's (the closest professional ground to me, and perhaps the most famous) to watch, sometimes with friends, sometimes with my grandfather, and sometimes on my own. One of my heroes at the time was one of England's most celebrated ever cricketers, Ian Botham. He was an all-rounder—one of that rare breed of player who is equally accomplished with both bat and ball—and was a tremendously entertaining cricketer to watch.

Some months ago, I received an invitation to go to a tasting of Ian Botham's new wines. Those attending would get a chance to meet Ian, ask him questions, taste his wines, and perhaps get a selfie or two. I didn't go.

Why? I've got nothing against Ian, and I still think he is one of the most exciting cricketers of all time. But I hate celebrity wines. To his credit, Ian does seem to like wine. He knows a bit about it, and he drinks it seriously. He takes part in the blending sessions for the wines with his name on the label, and probably sits in business meetings where the brand strategy is devised. The wines are made in conjunction with Paul Schaafsma, previously of McGuigan and Accolade. Paul is a good businessman, and this project has all the hallmarks of a good business opportunity. Botham claims that there is no celebrity in this project and that it's all about the wines. I'm totally uninterested.

Of course, there are far, far worse "celebrity wines" out there. (Indeed, as celebrity wines go, Botham's is probably one of the better projects.) The very worst is where a celebrity attaches their name to an unremarkable commercial wine. It's a simple endorsement, and I have no idea why, for example, people would buy an inexpensive Kiwi Sauvignon just because it has Graham Norton—a talented chat show host, but a winemaker?—on the label.

Often, celebrities need somewhere to park large sums of money, so they buy beautiful properties, some of which are wine estates. Sting, for one, has a swanky place in Tuscany. Wine is made from this estate. I've tasted it and it is good: typical, satisfying, made from organically farmed grapes. But I can't imagine Sting operating the press, or supervising the pick, or filling barrels. A genius musician, but he's not the reason to buy the wine.

If you are going to put your name to a wine, it's not enough just to be mates with someone in the wine business who can offer you a commercial opportunity, or to be rich enough to buy a wine estate. You need to know something about wine. You need to be the one making the calls. You need to know your vineyards, and you need to be in the cellar during vintage (or at least have spent some time in a wine cellar, and know what's going on there, so you can make the calls).

Celebrity wines. Yawn. No thanks.

52

All can play

*The wine world needs new talent and we
need to make sure we are creating a culture
that's open, where all can play.*

THE WINE WORLD needs new talent. It needs new
voices. It needs to reach out and draw in people from
diverse backgrounds who are genuinely interested in
wine. Indeed, the industry has a diversity problem,
which isn't helped by the increasing focus on difficult,
expensive exams that are out of reach for many unless
they have an employer who is willing to help defray the
cost.

In any profession, professional bodies spring up
that are intrinsically protective, restricting entry so
that their members have more work and get better
paid for it. But this conflicts with the desire to train

and develop new entrants and uncover new talent. Credentialism can be a problem.

It is a human tendency to raise the bar when we have already crossed it. Think back to the beginning of your career. Now think about your attitude toward those who are now in the position you were in then. Are you expecting more of those people than was expected of you when you started out? Are you allowing others to grow through experience, or are you insisting everyone has it all to start with? Raising the bar behind you only restricts entry to new talent.

No industry thrives when second-rate people clog up the nice jobs by virtue of having passed some tricky exams, some of them many decades ago. It doesn't help that many of these exams have a distinctly British focus to them, because these exams, open to members of the wine trade globally, are designed by UK-based educational bodies. These days the wine world is big, and global, yet it could be argued that to contribute in an excellent way shouldn't require (a) a global knowledge of wine (which is hard to get unless you are based in a market that has a global set of wines represented, and there are few of those) or (b) a peculiarly British slant on the world of wine.

Education is important. Let's not use it as a barrier to entry to good people who through their circumstances can't study as easily as others can. Credentialism is horrible.

53

How to succeed at wine writing by writing boring articles

Most writing in wine magazines is boring and formulaic. That seems to suggest that if you want to be a successful published wine writer, you need to learn how to write boring, formulaic articles, which is clearly what the market is looking for.[1]

YOU WANT TO be published? I'm here to help. In this chapter, I'll tell you the secrets—give you the inside track—on how to write an article the wine magazines will go for.

First of all, you need to take a press trip. Two or three days in wine region X, paid for by a generic body, where you get to visit a mix of producers. Traveling

---

1. I'm pretty sure this footnote is not needed, but you never know. So I should point out here for those who are wondering whether I am being sincere or not: this chapter is written in a spirit of satire, trying to point out, using humor, both the absurdity of some of what passes as wine writing and the fact that this is a broken system. I published it on my blog, but I thought it would fit perfectly in this book.

with a group of fellow writers, you'll be taken to one or two boutique producers, one or two larger producers, and some lousy huge producers who pay a lot of money to support the generic body. The exact itinerary, of course, will mostly be determined by internal politics. (Bad producers, you see, don't realize that it would be better for them if journalists just visited the best producers in any particular region.)

You then get a commission from a magazine editor to write about wine region X. It works out nicely, because if region X has the money to host journalists, they'll also have money for advertising. They will want 1,500–1,800 words, and they'll pay you between £225 and £250 per 1,000 (a rate that has not budged in fifteen years).

So how do you write your boring wine article? You haven't got room to go into any depth, so remember: big overview without many specifics. The good news: it won't take long to do, especially if you follow my template here.

Start off with how thirty years ago region X wasn't making very good wines, despite the obvious potential of the vineyards and the grape varieties grown there. Then explain the work of the pioneers. People who began making slightly better wines than their peers. Mention between one and four producers who found out that if they made better wines, they could charge more for them, and how they realized they were heading to the top when they won a trophy at a competition or got a 90+ score from Robert Parker.

Then put some facts in. How many hectares? Which varieties? What's the climate like?

Two paragraphs in, begin inserting a few quotes from some of the producers you visited. The blander and more generic, the better. Keep it positive.

Say how some producers are small, some are medium-sized, and some are big. Remark on how good the quality of the big producers is, considering how big they are.

Talk about the viticulture and winemaking. Explain that some of the vineyards are new plantings, whereas others are older. Explain that the best wines come from old vines on the best terroirs. Point out that some terroirs are better than others, and how the range of soil types varies across the region. You'll need to include a quote from one of the people you visited describing their soils, because you don't know what the geological terms they are using mean. (It's OK. No one does.) Mention the grape varieties that are grown here. Some of them are white and some are red.

There are different clones, and some of the clones are better than others. Some wines are varietal and some are blends. Sometimes blends are better than varietal wines, but sometimes they are not. Some producers use oak and some use stainless steel. Some use both.

Some producers are traditionalists and some are modernists. Some people think the traditionalists are right, but others think the modernists are better. Some producers are traditional, but in a modern sort of way.

Take a personal angle. Tell (briefly) the story of winegrower Y, who had a passion for wine and so bought a small vineyard and lavished it in passion, then gradually grew their business by making slightly better wines each year, all the while gradually increasing their vineyard holdings because of their passion for wine. Also explain carefully that some old winemakers are retiring and new winemakers are starting out.

But don't deflect from your narrative theme, which is this: everything is getting just a little bit better.[2] The wines being made today are better than

2. I think I have UK wine writer Richard Neill to thank for this insight, but I can't find the original quote.

those being made a few years ago, and because everyone is so passionate and motivated, we can confidently predict that things will continue to improve, little by little.

At this stage you'll be worrying that your article is beginning to read like an advertorial. (Which it is.) So you'll need a crunch. This is the time to introduce two or three mild threats or challenges to the ever-improving wine quality, but it's best that these challenges are fairly easily surmountable, or—even better—have already been successfully overcome. They could include a recent vintage that wasn't quite as good as an earlier one, or the risk of hail, or exchange rate instability, or the shortage of donkeys for old-style vineyard work.

To finish, the conclusion: The future is bright, and the wines from region X are better than ever. And you should probably be buying them.

This is where you recommend five producers. Include a range in terms of the scales of operation and quality levels you saw, picking a wine from each and giving it a vague generic description. Job done! No need to thank me; just trying to help.

54

The importance of stories

It's an unfortunate truth: Facts don't change people's minds. But stories do. Stories have an ability to weave themselves into our thinking and create change in a way that facts, however well presented, rarely do.

ONE OF THE most popular programs on BBC Radio 4 is *Desert Island Discs,* presented by Kirsty Young. It has been running forever (well, since 1942). The show comprises an interview of a notable person, with the broadcast punctuated by eight pieces of music they'd want to have with them if they were ever to be stranded on a desert island. I recently caught the episode with writer Ali Smith.[1] One interchange in particular struck me:

SMITH: Stories are incredibly powerful. We think we live, we're just living along going from day to day. Actually, we

---

1. This program was aired November 11, 2016, and can be heard at www.bbc.co.uk/programmes/b081tflr.

live by telling ourselves stories about the lives we are living. We take in, like sponges, the stories that come at us on all the waves—on all the radio waves, the TV waves, the internet—everything is a kind of story, which all adds to the story which is supposed to be the story of each individual's life. So it is not surprising that if the stories are good, and they come at us and we are the sponges that take the stories in, then we will feel better about it. And those stories are coming at us, and us being so porous, if we aren't careful with our stories then we will probably block our pores.

YOUNG: If there are right stories, then by definition there are wrong stories that can do harm. That seems quite a curious thing for a writer to say.

SMITH: We have to know that our lives are narrated to us, and also the way that we narrate lives around us. It is all construct. As soon as we become aware of that, we can do whatever we like with the construct: we can change it if we need to, we can stay with it if we like it, we can change bits of it. In other words, it gives us a kind of empowerment. So if we are not careful, [stories] will take the shirts off our back, but if we are careful, the stories will see us through like boats on whatever surface the sea is doing.

The idea here is one I've been interested in for a while. It is stories that mold us, and if we want to change, then we need to reformulate these stories—

or integrate fresh ones. You can't change people's minds by presenting them with facts. You have to use stories.

There is very little that is neutral in our culture. Most information we are exposed to comes with narrative attached. As Smith points out, we are like story sponges. We soak up narrative and it becomes part of us.

Our culture is full of stories, and inevitably, we pick them up. As I write, in February 2020, the UK has just left the European Union, following a referendum almost four years ago. For many of us it is hard to see how anyone could have found the claims made by supporters of Brexit to be plausible. Yet a certain proportion of people in the UK have been soaking up very different stories, and their belief that leaving the European Union is good for the UK is a result of sponging up what many readers will regard as "bad" narratives.

I'm interested in stories and how they apply to wine. I think there are similar narratives in wine that influence how we feel about certain regions and styles of wine. A great example of wine storytelling is Kermit Lynch's excellent *Adventures on the Wine Route:*

A Wine Buyer's Tour of France (1990). This narrative has, I think, had a strong impact on the world of wine, particularly for those Americans who know of Lynch and have read the book. The quest for real, authentic, and natural wine is, remember, a relatively recent one.

I like to think we are moving away from the points system. Wine is so diverse, and the idea that a wine can be summed up in a simple score is absurd (although granted, a score of some sort can help readers know how much you liked the wine). More telling, surely, is the story that has at its center the notion that a wine is of a place. I buy a Chablis and celebrate its Chablis-like essence. A good Chablis is one that is a sensible, skilled interpretation of that place.

To suggest that the merit of a wine lies in how much you "enjoy" the flavor, or how much hedonic appeal it has, is nonsense. If you view the pinnacle of perfection simply as 100 points, place is reduced to merely a means of helping create this perfect wine, and not as something important in its own right.

Do wine journalists matter? I think so, and not just because their recommendations affect the sale of specific wines. It's because good journalists tell stories, and this narrative then shapes how people feel about regions, varieties, producers, and vintages. Just as Ali Smith points out, there are good narratives and bad ones. As journalists, and as a wine trade, we need to be careful that we are telling the right sorts of stories.

55

Why it matters

Wine matters because it frees us to share;
it brings us together; it leads us to truth;
it celebrates life.

THIS IS THE last chapter of this short book, and it's
where I make a case for wine. For many, it is merely
a drink, containing enough alcohol to mellow mind
and body. But should we want to pose more questions
to the wine in our glass, we find that there can be a
lot more to this drink. It repays attention. It's like the
friend who you enjoy hanging with, but don't attribute
much to in the way of smartness and personality—until one day you find yourself in a situation where they
surprise you: they have hidden depths that you were
unaware of. They have substance.

At its heart, wine is a sociable drink. It is not alone
in this regard, but the particular association of wine

with the table cements its position at the heart of that most intimate of shared social moments: mealtime. In traditional wine-producing countries, wine was frequently served with food.[1] So attractive is the culture of wine, that it has spread to countries where wine has not been a traditional drink.

The privileged place of wine at the table has continued with the restaurant experience. We sit down and are handed a menu and a wine list. The wine list may contain other drinks, but primarily it focuses on wines.

The planting of vineyards was one of the first acts of settlement, at least in countries of the vine. Here was a gesture that said, I am staying here, this is my place. Where the grapevine was grown, wine became a central part of community life. It even gained religious significance, at the center of the Christian worship along with bread. It is a celebratory drink, too, especially when it has bubbles.

---

[1]. In traditional wine-producing countries, wine often accompanied agricultural workers into the field. In Portugal, a favored style for this work-drink was Palhete, a blend of red and white grapes, low in alcohol and body, vibrant in color, and ideal for quenching thirst. These smashable, refreshing wines are now coming back into fashion, having almost gone extinct.

Wine also speaks of time. One of the things we must recognize in life is that time never stops, and we can live in—and cherish—the moment, but we can never stay in it. We are born, we grow up, we live, and we die, and we pass the baton on to future generations. But wine is able to transcend time. Drinking an old vintage transports us to the year that wine was made. And the planting and cultivation of a vineyard is an act of faith, looking forward to the future. In many cases, vineyards are planted or renewed for the next generation. There is a wisdom in wine, just as there is wisdom in planting a tree.

Wine has a habit of grabbing people and not letting go. More so than any other drink, it engages people's intellects. It becomes a hobby for many, an obsession for a few. When people make a lot of money, one of the things they like to do is collect wine, or even buy a vineyard.

As well as connecting us to a time, wine connects us to a place. Almost all interesting wines are inseparable from their vineyard origin, whether broadly applied (the region) or more precisely dialed in (single vineyards or parcels). The exact way this place is conveyed is impossible to define. But in the differences among wines we find this truth of terroir. Some of the communication of place is human-inspired—indeed, it is hard to remove humans from terroir, even if it would be tidier that we did.

I could go on. It's enough to say that wine is worthwhile, complex, and repays attention. It is special. We need to cling to the cultural richness that has its origin in a time and place, and celebrate this fabulous gift. That's why I wrote this book.

Acknowledgments

I OWE A debt of gratitude to all those who have helped
and supported me on my wine journey. To the edi-
tors who have commissioned work from me; to the
PR professionals who have helped connect me with
producers and wines; to the regions and winegrowers
who have been hospitable, open, and tolerant in letting
me explore their vineyards and wineries; and to my
peers, who have shown friendship and encouragement.
But there are also those I have never met who have
supported me by buying my books, following my social
media accounts, and visiting my website. I don't take
them for granted. We exist in community, intercon-
nected. That's what is great about the wine world.

Index

Founded in 1893,
UNIVERSITY OF CALIFORNIA PRESS
publishes bold, progressive books and journals
on topics in the arts, humanities, social sciences,
and natural sciences—with a focus on social
justice issues—that inspire thought and action
among readers worldwide.

The UC PRESS FOUNDATION
raises funds to uphold the press's vital role
as an independent, nonprofit publisher, and
receives philanthropic support from a wide
range of individuals and institutions—and from
committed readers like you. To learn more, visit
ucpress.edu/supportus.